GENERAL EQUILIBRIUM AND GAME THEORY

GENERAL EQUILIBRIUM
AND
GAME THEORY

Ten Papers

Andreu Mas-Colell

With an Introduction by Hugo F. Sonnenschein

Harvard University Press

Cambridge, Massachusetts
London, England
2016

Printed in the United States of America

First printing

Library of Congress Cataloging-in-Publication Data

Mas-Colell, Andreu.
[Essays. Selections]
General equilibrium and game theory : ten papers / Andreu Mas-Colell ;
with an Introduction by Hugo F. Sonnenschein.
pages cm
Includes bibliographical references.
ISBN 978-0-674-72873-8 (alk. paper)
1. Equilibrium (Economics) 2. Game theory. I. Title.
HB145.M3797 2016
339.5—dc23
2015017577

Contents

Preface

This is a book that would not exist without friends. They have taken the initiative to produce it, and they know well enough that they had to push me to collaborate properly. So I can only start by thanking them: Hugo Sonnenschein, Antoni Bosch-Domènech, Xavier Calsamiglia, Joaquim Silvestre and, more in the background, Jerry Green. I must have done something right to have such friends.

In the process I have learned the hard way why people like to publish their complete collection of papers. It saves on the torture of having to select. It is already very difficult if one has to do the selecting on someone else's work. It is excruciatingly difficult to do it on one's own work. How have I done it? Since the selection had to be severe, I decided to focus on the two general areas of research in which I have concentrated during my career: general equilibrium and game theory. This has left out some papers that I hate to have left out; one is my joint, very early, paper with Hugo Sonnenschein in Social Choice Theory. I did not pursue this line of research, and I have some regret for it. In addition, I have paid attention to the "market," that is, to citations and the perception of impact. But not exclusively. I have not resisted rescuing some (few) papers of which it is clear that I am more fond than the market is. Maybe I want to give them another chance by showing my preferences. This said, I hate the idea that my selection could contribute to sending my unselected papers, or some of them, to the realm of oblivion. Yet the rational part of my mind tells me that, at the end, the scholarly community will place each of my papers, selected or unselected by me, in its proper place, whatever that may be.

I have been thought often to be a mathematician turned economist. But this is not so, my background is as an economist. I became an economist for reasons common to so many others: it appeared to me, and still does, as the right intellectual tool to improve the lot of the people, advance justice, and modernize society. My undergraduate training, at the University of Barcelona, was rich in exposition of economic doctrines, in institutional detail, in law, in history (I read my Adam Smith, Ricardo, Marx, and Keynes at an early age) . . . but short in analytics. That I evolved towards analytics is probably a testimony either that I was made for this or that I was not made for the more descriptive approach. That once in the analytical stream I evolved towards microtheory is probably a combination of the empirical approach being too difficult for me (Max Planck made this remark when explaining why he did not pursue economics and chose physics, and I'm making it in exactly the same spirit), of the influence of teachers and mentors, and, I have to hope, of the genuine interest of the challenges thrown.

I'm of the kind who become fascinated by what they are not good at doing. But I do not regret (in fact, from youth I committed not to turn with the years into a critic of an earlier me) having followed my comparative aptitudes, such as they may be, towards theory, theory being an integral part of scientific economics. Not always does the development of theory and empirics go in parallel in economics. There are times where theory is in the limelight, and trendy, and times where it is the turn of empirics. This is not merely because of herding effects. There is also a natural cycling component. After a period of intense theoretical work, it is inevitable that a sentiment of getting lost in the clouds develops and that an irresistible urge to touch ground emerges. In turn, after an intense phase of empirics, the craving is for understanding, which leads to the appetite for theory. At any rate, be as it may, I was fortunate: I arrived at the scene in a phase of ascendancy of theory, which fitted my tastes and aptitudes.

A word about my teachers and mentors. At the risk of unjustly leaving out some names (I leave out my generational peers), I would like to mention M. Sacristán, J. Nadal, and F. Estapé from the time of my undergraduate studies in Barcelona. J. L. Rojo was the center of my first postgraduate studies in Madrid. Then came Minnesota, which definitely shaped the direction of my career (and where I learned my mathematics). The names there were H. Sonnenschein (to whom I owe so much), L. Hurwicz (to whom I should have paid more attention), and M. Richter (who directed my Ph.D. thesis). I started as a postdoc in Berkeley in an intellectual atmosphere dominated by Gerard Debreu (I should also mention R. Radner and D. McFadden), who had a major impact on my research. My heartfelt thanks to all of them.

Let me put on record that I'm overwhelmed by the Introduction that Hugo Sonnenschein has written for this volume. It is not something I could expect, and it is yet another instance of a well-known fact: his extreme generosity. If one could blush on paper, then I would be blushing here. Thanks, Hugo. I also smile at the title, and content, of the biographical appendix by my dear friends, colleagues, and comrades-in-arms of many battles: Antoni Bosch-Domènech, Xavier Calsamiglia, and Joaquim Silvestre. Thanks, guys.

In his Introduction Hugo has commented on six of the papers published in this volume. Allow me now a brief note on each of the remaining four.

"Efficiency and Decentralization in the Pure Theory of Public Goods" (Chapter 4 in this volume). This paper belongs to the category of those selected with the intention to give them a little push. It is at its core a paper on the First Welfare Theorem. What makes the theorem tick? Leaving aside issues of market completeness, the standard theory gives the following condition: linear prices that are identical across agent. So general a result deserves to be true even if the commodity space lacks any linear structure. And indeed it is. An appropriate formulation is provided in this paper in terms of "valuation functions." Incidentally, a little thinking will reveal that nonlinear prices identical across agents will not do.

"The Price Equilibrium Existence Problem in Topological Vector Lattices" (Chapter 5). This is a paper on unification and abstraction. Equilibrium theory with infinitely many commodities is more general than, say, the finitely many commodities of G. Debreu's *Theory of Value* in precisely this respect: infinite versus finite. But in every other respect it is more restrictive, because conditions are needed on the structure of the commodity space. This is easy to understand: *finite* dimensional means the n-dimensional Euclidean space, but *infinite dimensional* does not have a unique meaning. The conditions are dictated by the particular economic problem at hand. It is not the same if we are trying to model consumption over time, returns of financial assets, or differentiated commodities. The seminal contribution of T. Bewley fitted time well, but less so financial assets or differentiated commodities, for which specific models had to be developed. In this paper it is shown that the mathematical structure of topological vector lattices, which had been used by R. Aliprantis, fitted very well the logic of Pareto optimality and allowed for an encompassing existence theorem.

"Potential, Value, and Consistency" (with Sergiu Hart, Chapter 7). This is the first product of a very long, and for me extremely fruitful, collaboration with Sergiu Hart. I have been most fortunate in this partnership, for the obvious intellectual reasons but also for the fact that Sergiu's discipline has kept me focused (at least in short, but very intense, meetings) on our research agenda even at times when I fall into distractions of all types, something to which I'm prone. On a more personal note, he has become a close friend, and, punctuated by frequent and happy visits to Jerusalem, the same has occurred for our families. The paper at hand looks at the Shapley value and provides a characterization at once novel and rich in implications: the Shapley value is the only assignment of imputations to every subgroup of players that it is "integrable," that is, that admits a *Potential*—a real-valued function on the space of subsets of the set of players with the property that, for each coalition, the Shapley value is the (difference rather than differential) gradient vector of the potential function at that coalition.

"An Equivalence Theorem for a Bargaining Set" (Chapter 8). It is a remarkable fact that solution concepts of cooperative game theory have turned out to be closely related to the Walrasian equilibrium outcome, both in their outcomes and when applied to exchange economies with many players. This was the case first for the notion of the Core and then for the Shapley Value. In this paper it is shown that, with appropriate and natural definitions, it is also true for the Bargaining Set. As is the Core, this concept (by R. Aumann and M. Maschler) is based on the dominance relation. But, in contrast to the Core, to block is much more demanding, because a blocking coalition can only do so if the blocking imputation is "justified" in the sense of not being itself blockable by the same logic. This amounts to a kind of internal stability or consistency requirement. It follows that the Core is smaller than the Bargaining Set, and it is thus a far-reaching generalization of the Core Equivalence Theorem to be able to show that the non-Walrasian allocations can be blocked by coalitions and allocations subject to further demanding conditions (for example, it turns out that the justified blocking imputation has to be Walrasian in the blocking coalition, although this by itself is not enough).

GENERAL EQUILIBRIUM AND GAME THEORY

Introduction

HUGO F. SONNENSCHEIN

The purpose of this volume is to honor the scholarly contributions of Andreu Mas-Colell. It collects ten papers on economic theory that were written by Mas-Colell over a period of thirty years. They were selected by the author and include his most frequently cited scholarly work. The subjects range from general equilibrium theory to foundational issues in finance and game theory. My aim in this Introduction is to explain Mas-Colell's place in modern economics, with particular reference to the papers included here. I will conclude with some recollections of his years as a student at Minnesota and the beginning of his time at Berkeley, and finally, I will speak briefly about his transition from economic theorist to scholarly leader and public servant.[1]

Like most formal theorists of his generation, Mas-Colell was profoundly influenced by the work of Arrow and Debreu and their contemporaries, who in turn benefited from the Hicks-Samuelson syntheses of Microeconomic Theory (Samuelson 1947; Hicks 1939) and the von Neumann-Morgenstern formulation of Game Theory (von Neumann and Morgenstern 1944). The work of several economists is particularly important in form and substance and leads most directly to Mas-Colell's work: Arrow and Debreu on general economic equilibrium (Debreu 1952; Arrow and Debreu 1954), Arrow for his axiomatic casting of the problem of social choice (Arrow 1951), Leonid Hurwicz, with whom Mas-Colell studied at the University of Minnesota (Hurwicz 1960), for his framing of mecha-

1. I am pleased to acknowledge the helpful comments of Michael Aronson, Salvador Barberà, Jerry Green, Sergiu Hart, David Kreps, Peyton Young, and particularly Wayne Shafer. I also wish to thank Carla Reiter for editorial assistance.

nism design, and Lionel McKenzie for his independent contribution to the general economic equilibrium existence theorem (McKenzie 1954).

Among these, the Arrow-Debreu Theory of general economic equilibrium was pivotal. It was immediately followed in the 1960s and early '70s by an outpouring of important contributions to which no short list can do justice. But it is important to mention the work by Scarf on computational methods for finding equilibrium (Scarf and Hansen 1973), Debreu and Scarf (stimulated by Shubik) on the Core (Debreu and Scarf 1963), followed by Aumann, and Vind, and later Hildenbrand, who in addition to furthering the work on the Core made precise the notion of an economy with a large number of infinitesimal agents (Aumann 1964; Vind 1964; Hildenbrand 1974). All of these contributed to a deeper understanding of the Arrow-Debreu Theory and were pivotal to the development of Mas-Colell's thinking.

Andreu Mas-Colell entered the University of Minnesota for his Ph.D. studies in economics in 1968 and completed his dissertation under the supervision of Marcel K. Richter. In the sixties and early seventies Minnesota and Berkeley were two places where the mathematical approach to economics held particular sway. So it was no surprise that Mas-Colell took his first position at Berkeley in 1972. His generosity to colleagues and their respect for the breadth of his knowledge led to increased responsibility in the graduate programs at Berkeley and then at Harvard, where he moved in 1981. Not only did Mas-Colell attract outstanding thesis students, but he was also very good at introducing entering doctoral students with a broad range of potential research interests to modern economic theory.[2] His pedagogical work of this period culminated in the text *Microeconomic Theory* (with M. Whinston and J. Green), which saw the light of day in 1995 (Mas-Colell et al. 1995). While this is a coauthored book that builds upon the research and teaching of many individuals, the influence of Mas-Colell is apparent and reflects his intellectual leadership. It is perhaps the most influential textbook for graduate economics in a more than twenty-year period: a worthy successor to the treatises of Hicks and Samuelson.

2. The following is a chronological list of those who completed their Ph.D. under Mas-Colell's supervision. At Berkeley: Hsueh-Cheng (Harrison) Cheng, Norbert Schulz, Nicholas Economides, Nirvikar Singh; at Harvard: Michael Mandel, Lars Tyge Nielsen, Mathias Dewatripont, John Nachbar, Michael Spagat, Andrew Newman, Atsushi Kajii, Roberto Serrano, Chiaki Hara; at the *Universitat Pompeu Fabra:* Margarida Corominas Bosch, Antoni Calvó-Armengol, Rasa Karapandza, Sandro Shelegia.

This is not simply because it contains so much of what we must know in order to speak with one another, but also because it is wise, deeply synthetic, and analytically state-of-the-art. It equips the reader with the theoretical knowledge to confront a broad range of important applications, which include, for example, industrial organization, labor economics, financial economics, and international economics.

The papers in this volume begin in 1974, and the early papers bear witness to the speed with which Mas-Colell became influential in the field. Over the twenty-year period from 1974 to 1994, he came to be known as one of the very most analytically powerful economists of his generation.[3] At the same time, he emerged as someone who had mastered a broad range of economic thinking, had excellent judgment, and was a leader in creating and synthesizing the next chapters of microeconomics learning.

The papers included here represent the best of Mas-Colell's scholarly contributions. More than any of his many other achievements, this is the material that places Mas-Colell as the worthy heir to Gerard Debreu, with whom he served at Berkeley in both the economics and the mathematics departments. This is high praise and I will explain why it is appropriate to view his contribution in this manner before turning to the individual research contributions.

Debreu believed strongly that a formal mathematical reworking of the Walrasian theory of value would play a major role in revolutionizing economics. His representation of the Arrow-Debreu model in his *Theory of Value* (Debreu 1959) was his crowning achievement, and built upon Debreu (1952) and Arrow-Debreu (1954). Coupled with Arrow's *Social Choice and Individual Values* (Arrow 1951), it established the power of the axiomatic approach to economics. It was also the place where a significant number of economists were introduced to mathematical tools that are now viewed as essential to modern economic theory. Debreu was a missionary; he believed and argued that the new approach, and the new tools that he and Arrow introduced, had led to a deeper understanding of fundamental issues in price theory and to notable gains in accuracy, generality, and simplicity. He believed in the long-term impact of the new approach on all of economics, and he embarked on a research program that was guided by these principles. He lived to see a world where one could not attend a se-

3. See, for example, *The Theory of General Economic Equilibrium: A Differentiable Approach* (Mas-Colell 1990).

ries of lectures in monetary economics, finance, or international trade without hearing the phrase "Arrow-Debreu model." Moreover, real analysis, convex analysis, dynamic programming, and measure theory have become standard elements of the economist's tool kit. From microeconomics to macroeconomics, and from the most theoretical to the more applied, by the standard of fifty years ago we are now all mathematical economists. But even for Debreu it may have been difficult to envision the substantial technical challenges that lay ahead in recasting the Walrasian theory to fit a rich variety of applications. Moreover, he could not have foreseen the extent to which theories of bargaining, auctions, and matching would integrate price theory and game theory and eventually lead to rich empirical and practical applications.

Mas-Colell has led in the advancement of Debreu's research program and point of view. The papers included here, as well as others not included, illustrate how Mas-Colell broadened the reach of the mathematical approach to include, for example, central questions in finance, industrial organization, and public economics. They also illustrate his influence upon method. As with the contributions of Debreu, Mas-Colell's papers are the references upon which to build, and because of their excellent craftsmanship and attention to the most basic issues, they will be with us for a long time.

The Papers

I provide here a commentary on some of the papers in this volume. The choice of which papers to cover is subjective and reflects my own particular interests and abilities and should not be interpreted as suggesting which are most valuable.

The first paper in the collection (Mas-Colell 1974) concerns the extension of the Arrow-Debreu Theory to include the possibility of agents who are less than "rational" in the specification of their preferences, a move that opens the door to interpretations that are "behavioral." Making clear the obstacles to achieving the goal of this paper requires some background. The normal rationality requirement for consumers demands that for any possible commodity bundles x and y it must be the case that x is at least as good as y or y is at least as good as x (completeness). Furthermore, x preferred to y, y preferred to z, and z preferred to x is not possible (an implication of transitivity).

In the absence of completeness, if a budget set contains only the two bundles x and y such that x is preferred to y and y is preferred to x, then the choice of either x or y is problematic. Similarly, if there is a budget set composed of x, y, and z with x preferred to y, y preferred to z, and z preferred to x, then the choice of x, y, or z is problematic. But of course, in the context of the other axioms of general equilibrium theory, both budget sets and the set of bundles preferred to any given bundle are convex, so these two problematic examples do not contradict the possibility of a general equilibrium theory with standard convexity. There were some hints in the previous literature that this might be manageable (Schmeidler 1969, and particularly Sonnenschein 1971, which was widely circulated by 1965 and studied by Mas-Colell), but Mas-Colell's paper quite simply put the problem to rest by first recasting the definition of preferences as a map from states of the economy to a consumer's preferred bundles, and then imposing convex-valuedness and suitable continuity on this map. This turns out to be the essence of what is needed for the general existence of equilibrium, and it frees the theory from preference relations, completeness, and transitivity. The enduring importance of Mas-Colell's contribution is manifest in the increasing attention that is given to explaining economic phenomena in which agents are less than perfectly rational.[4]

The second paper in the collection (Mas-Colell 1975) concerns the extension of the Arrow-Debreu model to the case of differentiated commodities, as in the pioneering but less than mathematically precise formulations of Chamberlin and Robinson (Chamberlin 1933; Robinson 1933) and the precise but mathematically narrow formulation of Sherwin Rosen (1974). Mas-Colell posits a continuum of substitutable commodities and a continuum of consumers. No consumer comes to the market with an amount of differentiated commodity that allows him to exercise market power. The paper builds mathematically on earlier work on markets with a continuum of agents. (Particularly from the point of view of the formalism, one should mention the contributions of Truman Bewley [1970], who also studied the case of equilibrium with a continuum of commodities at approximately the same time.) Mas-Colell's paper is an analytical tour de force. It requires the full power of a continuum of both commodities and agents. At the time it was written, there was likely only a handful of economists who possessed the analytical power, not to mention the modeling

4. With the benefit of hindsight, one sees that the original attack of Arrow and Debreu on the Existence Theorem via generalized games has some substantial advantage (Debreu 1952; Arrow and Debreu 1954), since their utility functions can be interpreted as representing the set of a consumer's preferred bundles for each state of the economy.

judgment, to pull it off. It delivered a model for economies with an infinite number of commodities that both allows you to prove the existence of equilibrium and gets to the heart of where equilibrium and the Core coincide.

As it turns out, the extension by Mas-Colell and Bewley of the Arrow-Debreu theory to economies with an infinite number of commodities has been particularly fruitful. Work that establishes a precise mathematical foundation for arbitrage pricing in finance (see, in particular, Kreps 1981) exploits Mas-Colell's treatment. Ideas of private information, adverse selection, and moral hazard became increasingly important in the economic modeling of monetary, financial, and labor market equilibrium. As they did, the greater generality and applicability achieved in the pioneering papers on economies with an infinite number of commodities were increasingly recognized. (See, for example, Prescott and Townsend 1984; Parente and Prescott 1994; Cole and Prescott 1997.)

The third paper here (Mas-Colell 1977) is particularly close to my heart, since it offers a substantial advance on the so-called Sonnenschein–Mantel–Debreu Theorem on the structure of excess demand functions (Sonnenschein 1972; Mantel 1974; Debreu 1974). The question at hand for Mas-Colell concerns the set of possible equilibrium price sets for an Arrow-Debreu economy: how do the assumptions of utility-maximizing behavior for consumers and profit-maximizing behavior for firms limit the set of prices that clear markets? This question is intimately related to the structure of excess demand functions, which the above-named authors solved with increasing generality for compact subsets of the open price simplex. Mas-Colell provided a refinement of Debreu's treatment of the excess demand function theorem that was sharp enough to extend the result to a large enough compact subset of the price simplex to characterize the equilibrium price sets. This is a delicate extension that requires some nontrivial differential topology. It stands as the definitive answer to a basic question in general equilibrium theory.[5]

The sixth paper, titled "Real Indeterminacy with Financial Assets" (Geanakoplos and Mas-Colell 1989), is joint work with John Geanakoplos. The starting point is Arrow's extension of the Arrow-Debreu model to in-

5. Theorems regarding the existence of general economic equilibrium guarantee that the equilibrium price set must be nonempty. Furthermore, the interpretation of the value of excess demand as one moves away from equilibrium prices is questionable. So, from the point of what can be stated about an economy and equilibria, it is the equilibrium price set that deserves special notice.

clude securities (called Arrow securities) that promise to deliver one dollar in a specified state and zero in other states. Arrow proved that when spot prices for these securities are correctly anticipated, equilibrium allocations are the same as in an Arrow-Debreu world with complete contingent claims. David Cass provided an important example of an economy with one financial asset and two states in which there is a one-dimensional continuum of real equilibria (with the interpretation of an infinite amount of "value indeterminacy") (Cass 1984, 1985). There is good reason to question the descriptive relevance of markets in which there are Arrow securities for each state of nature, especially when there is asymmetric information. So Cass's example and the work of others have led to a great deal of interest regarding the nature of indeterminacy, perhaps to be thought of as the basis for a theory of bubbles. Mas-Colell and Geanakoplos proved the surprising result that when there are fewer Arrow securities than states, the dimension of indeterminacy is $S - 1$, where S is the number of assets and thus independent of the number of Arrow securities. In their own words, "let just one financial asset be missing and the model becomes highly indeterminate." This is certainly one of the landmark papers that extend the Arrow-Debreu model to include uncertainty, and it is among the very most cited in the important strand of the literature that studies the consequences of there being an incomplete set of Arrow securities.

The final two papers are joint with Sergiu Hart and concern learning in noncooperative games. They were written after Mas-Colell's return to Spain and during a time when he had become increasingly occupied with public service. He simply made the time for this most important collaboration. The papers concern the general question of dynamic adjustment processes for games. Just as general equilibrium theory is "incomplete" without a story of how and why one may find one's way to Arrow-Debreu equilibrium, the theory of noncooperative games calls for descriptions of how players find their way to Nash equilibrium and correlated (Nash) equilibrium. There is some difference of opinion regarding whether Nash equilibrium or correlated equilibrium is the natural way to conceive of a solution for noncooperative games. There are sensible defenses for each position. However, in the absence of sensible dynamics that get a social system to its equilibrium, these concepts are at the very least incomplete.

In their paper "A Simple Adaptive Procedure Leading to Correlated Equilibrium" (Chapter 9 here), Hart and Mas-Colell (2000) put forth a simple adaptive procedure and use an important theorem of Blackwell to demonstrate that it always converges to correlated equilibrium. This is not the first paper to propose a dynamic process that leads to correlated equilibrium, nor is it the first use of the central technique in this context. However, it is a particularly simple and elegant process. It also has the characteristic that it is "adaptive" or "behavioral" in the sense that a player's strategy does not depend on the utility functions of the other players. It may depend on the strategies of other agents, but not on the utility functions of other agents. Mas-Colell and Hart refer to such dynamics as "uncoupled" and note that in the world of mechanism design, this requirement has been referred to as "privacy preserving."

This sets the stage for the tenth paper, also with Hart, titled "Uncoupled Dynamics Do Not Lead to Nash Equilibrium" (Hart and Mas-Colell 2003), which shows that there are no uncoupled dynamics that guarantee convergence to Nash equilibrium. This is an extremely powerful result that presents an important challenge to Nash equilibrium as the central solution concept for noncooperative games. It is also important to note that Hart and Mas-Colell's impossibility theorem does not depend on the rationality requirement for players; it follows from the "framework" (perhaps in particular limitations on the state variable) and the informational requirement of uncoupledness.

I will point here to some of the themes that unify the papers in this volume, and in particular the ones that I have spoken about. Many of the papers concern basic extensions of the Arrow-Debreu Theory of Value, which has enabled modern price theory to include realistic elements of fundamental importance: behavioral agents, differentiated products, financial assets, and incomplete markets leading to theories of asset bubbles. Second, they have been fundamental to our understanding of some of the limits of both the Arrow-Debreu model of equilibrium and the noncooperative model of Nash equilibrium. The product-differentiation paper and some of the papers not explicitly considered in my comments give support to the Arrow-Debreu equilibria from the point of view of cooperative game solution concepts. Last, the third and the final two papers question our ability to conceptualize the premier notions of equilibrium in economics as rest points of economically attractive dynamic processes.[6]

6. This point is stated explicitly in the final paper in this collection. See point IV (b). Observe that the nature of uncoupled dynamics in general equilibrium is much better understood once one has in hand Mas-Colell's refinement of the Sonnenschein–Mantel–Debreu Theorem.

Concluding Remarks

This volume was conceived in April 2009 at Andreu Mas-Colell's sixty-fifth birthday celebration in Barcelona. Particular credit belongs to his college friends Antoni Bosch-Domènech, Xavier Calsamiglia, and Joaquim Silvestre. Some remarks at that celebration form the basis for my final words, which I acknowledge are less about Andreu's scholarly contributions and more about his personal life and achievements. In the spirit of that happy gathering, I will call the celebrant Andreu rather than Mas-Colell. My only advantage in preparing these remarks is that I bore witness to Andreu's remarkable growth as an economic theorist during the very early years at the University of Minnesota and then (less directly) as an Assistant Research Economist at Berkeley. This rapid growth paved the way for his advancement from assistant professor of economics and mathematics at Berkeley to full professor of both in just four years. At the end of that time his position as "heir to Gerard Debreu" was well understood.

I will not write from the perspective of teacher because, in truth, I have learned at least as much from Andreu as I have taught him. We shared an outstanding environment in which to learn at the University of Minnesota; the university was an excellent place to study the mathematical approach to economics. Leonid Hurwicz and John Chipman were distinguished senior members of the faculty, and Ket Richter had recently completed his groundbreaking work on revealed preference. My purpose here is simply to document some early impressions and to reflect upon Andreu's transition to builder of institutions and to public servant.

My first impressions of Andreu were of an individual with broad interests, a restless mind, unusual powers of persuasion, and a deep attachment to Catalonia and Spain. The young man who came to Minnesota in 1968 did not appear to have a stronger background in mathematics than most of the other graduate students of his time, however my experience with other Catalan graduate students suggests that they had a great aptitude for and interest in a mathematical approach to economics. But who knows, even this may have been an early Mas-Colell effect. In any case, Andreu was in no sense a mathematician by graduate or even undergraduate training, and this should be contrasted to the formal training of Gerard Debreu, Harald Kuhn, Herbert Scarf, Robert Aumann, David Gale, and Werner Hildenbrand (to make a point with some truly exceptional cases!). So it is hardly conceivable that his notes on the differentiable approach to economics (a precursor manuscript to *A Theory of General Economic Equilibrium: A Differentiable Approach*) were completed less than eight years after he entered graduate school: they were the basis for his spring of 1976 course in the Berkeley mathematics department.

I cannot claim to have seen immediately the unusual combination of abilities that went into his early achievements: extraordinary aptitude for mathematics and economic thinking, and an unusual capacity for work. In fact, beyond a general impression of "intensity," it did not occur to me that Andreu was especially hardworking and devoted to his studies—certainly not to the exclusion of all else. He always appeared to have time for friendships and politics, and sometimes these seemed intertwined. *The New York Times* and "the latest news from Spain" were always by his side.

It is not easy for me to account for the scholarly achievement that I have spoken about here, in particular, how quickly so much of it happened. Brilliance, capacity, and hard work surely play a role. But there must be something else, and in this regard it is useful to recall a statement of Andreu's daughter, Eva, who understands her father so well. She told us that her father "speaks a lot and knows about everything because he listens to everything." Andreu is a brilliant listener and a brilliant learner. This was an essential part of his mastering the mathematics that he has employed so creatively in such a short time. The breadth of his knowledge is equally impressive.[7]

Andreu doesn't merely listen; he listens well and he gives the impression of listening well. This has no doubt played a part in his success as a scholar, including in his work preparing the graduate text (Mas-Colell et al. 1995), which synthesizes such a broad variety of thinking, and in the education of graduate students. It has also likely played a role in his success in administration, in politics, and in the creation of educational and scientific institutions. We appreciate being led by someone with an attentive ear.

I recall presenting Andreu with a dilemma early in our relationship. He was enrolled in a course at the University of Minnesota in 1970 during the Vietnam War. There were protests against the war, and the student leadership called for the students to "strike" by not attending class. The strike was not sanctioned by the university, and professors (I was one) were expected to hold class. I recall Andreu negotiating a deal with me: rather than com-

7. When my wife, Beth, inquired how she might get information on the Sardana, a Catalan folk dance form, we were quickly directed to Andreu. And do not get him started on the history of the *barri gòtic*. He must ration his time somehow, but he also has great capacity.

ing to class and breaking the strike, he would write a paper on one of the several research projects that I had suggested during the course. The result was a joint publication in the *Review of Economic Studies* (Mas-Colell and Sonnenschein 1972). This was Andreu's first publication (and in an insecure moment I would argue that it deserves a place in this volume!). More to the point, not only was this my first view of Andreu's creativity, but it should be regarded as my introduction to Andreu as a master politician and teacher. I did not like the idea of students missing classes that I was expected to teach, and Andreu had to prove to me that the strike was not merely an excuse to miss class. I do not recall our negotiation, but it is highly unlikely that the compromise we arrived at was my idea. Andreu, as an active participant in Spanish resistance politics, was far more experienced in such matters and wiser about them than I.[8]

I will close with a recollection that is somewhat delicate to discuss. Despite Andreu's extraordinary devotion to economics and his quick success in the United States, I came to feel rather early on that he was on loan to the United States and even on loan to academic economics during his years at Minnesota, Berkeley, and Harvard. The delicacy of speaking about this belief comes from the fact that this is a book that will primarily be read by academic economists; many look to Andreu as a model, many have learned from him, and many could not imagine a more worthy life than the one that Andreu lived as a professor at Berkeley and Harvard. None of what I write here is intended to diminish that conclusion. Yet, in the richest lives there is time for different pursuits, and I tend to believe that Andreu was drawn back home by some of the same qualities that accounted for his academic success. Andreu is brave and not intimidated by the challenge of confronting new ideas. He listens well. He believes in the practical applica-

8. Our second paper together (Kihlstrom et al. 1976) was written in the summer of 1972 and was Andreu's first publication in *Econometrica*, the journal of which he subsequently became editor. This was the product of some nice weeks together in Amherst, Massachusetts, supported by the National Science Foundation. The event brought together Dan McFadden from Berkeley, Rolf Mantel from Instituto Di Tella in Argentina, Richard Kihlstrom and Leonard Mirman, who, with me, were members of the University of Minnesota faculty, and two very promising graduate students, Oliver Hart, who was recommended by Michael Rothschild at Princeton, and John Roberts, who had come with me from Minnesota. It is also noteworthy that during that summer Andreu did his first work on market excess demand functions. Andreu had only recently started at Berkeley. Suffice it to say that his leading role in the efforts of this group was a testament to his extraordinary growth during the early period.

tion of knowledge. He is tireless. He is a rigorous thinker. When one couples these qualities with his love of Catalonia and Spain, the possibilities for influencing change in Catalonia, and a strong sense of responsibility, it is perhaps not so surprising that he left Harvard at the peak of his academic career to create and lead institutions in Catalonia, Spain, and Europe. From his role in the creation of the *Universitat Pompeu Fabra,* to his position as the Commissioner for Universities and Research of Catalonia, from his position as Secretary General of the European Research Council, to his present position as Minister of Economics and Knowledge in the Catalan government, Andreu has been hard at work on activities that support the public good and for which he is particularly well suited.

We are grateful to have had Andreu "on loan." Thank you to Catalonia; to Andreu's wife, Esther; and to their exceptional children for sharing Andreu with us. We are grateful for the time he is able to spend on economic theory and in the support of economic institutions. We even maintain a hope that his current work on matters of great practical importance will lead to new perspectives and further results for economic science. We know how much he enjoys that work, and we want Andreu to know how much we look forward to learning more from him.

References

Arrow, Kenneth J. 1951. *Social Choice and Individual Values.* New York: John Wiley & Sons.

Arrow, Kenneth J., and Gerard Debreu. 1954. "Existence of an equilibrium for a competitive economy." *Econometrica* 22: 265–290.

Aumann, Robert J. 1964. "Markets with a continuum of traders." *Econometrica* 32(1): 39–50.

Bewley, Truman. 1970. *Equilibrium Theory with an Infinite-Dimensional Commodity Space.* Berkeley: University of California Press.

Cass, David. 1984. "Competitive equilibrium with incomplete financial markets." CARESS Working Paper No. 84-09.

———. 1985. "On the 'number' of equilibrium allocations with incomplete financial markets." CARESS Working Paper No. 85-16.

Chamberlin, Edward. 1933. *The Theory of Monopolistic Competition.* Cambridge, MA: Harvard University Press.

Cole, Harold L., and Edward C. Prescott. 1997. "Valuation equilibrium with clubs." *Journal of Economic Theory* 74: 19–39.

Debreu, Gerard. 1952. "A social equilibrium existence theorem." *Proceedings of the National Academy of Sciences* 38(10): 886–893.

———. 1959. *Theory of Value: An Axiomatic Analysis of Economic Equilibrium.* New Haven: Yale University Press.

———. 1974. "Excess demand functions." *Journal of Mathematical Economics* 1: 15–21.

Debreu, Gerard, and Herbert Scarf. 1963. "A limit theorem on the Core of an economy." *International Economic Review* 4(3): 235–246.

Geanakoplos, John, and Andreu Mas-Colell. 1989. "Real indeterminacy with financial assets." *Journal of Economic Theory* 47: 22–38. (Chapter 6 in this volume.)

Hart, Sergiu, and Andreu Mas-Colell. 2000. "A simple adaptive procedure leading to correlated equilibrium." *Econometrica* 68: 1127–1150. (Chapter 9 in this volume.)

———. 2003. "Uncoupled dynamics do not lead to Nash equilibrium." *American Economic Review* 93: 1830–1836. (Chapter 10 in this volume.)

Hicks, John. 1939. *Value and Capital: An Inquiry into Some Fundamental Principles of Economic Theory.* Oxford: Clarendon Press.

Hildenbrand, Werner. 1974. *Core and Equilibria of a Large Economy.* Princeton: Princeton University Press.

Hurwicz, Leonid. 1960. "Optimality and informational efficiency in resource allocation processes." In *Mathematical Methods in the Social Sciences,* ed. K. Arrow, S. Karlin, and P. Suppes, 27–46. Stanford: Stanford University Press.

Kihlstrom, Richard, Andreu Mas-Colell, and Hugo Sonnenschein. 1976. "The demand theory of the Weak Axiom of Revealed Preference." *Econometrica* 44 (5): 971–978.

Kreps, David. 1981. "Arbitrage and equilibrium in economies with infinitely many commodities." *Journal of Mathematical Economics* 8(1): 15–35.

Mantel, Rolf. 1974. "On the characterization of aggregate excess demand." *Journal of Economic Theory* 7: 348–353.

Mas-Colell, Andreu. 1974. "An equilibrium existence theorem without complete or transitive preferences." *Journal of Mathematical Economics* 1: 237–246. (Chapter 1 in this volume.)

———. 1975. "A model of equilibrium with differentiated commodities." *Journal of Mathematical Economics* 2: 263–295. (Chapter 2 in this volume.)

———. 1977. "On the equilibrium price set of an exchange economy." *Journal of Mathematical Economics* 4: 117–126. (Chapter 3 in this volume.)

———. 1990. *The Theory of General Economic Equilibrium: A Differentiable Approach.* Cambridge, MA: Cambridge University Press.

Mas-Colell, Andreu, and Hugo Sonnenschein. 1972. "General possibility theorems for group decisions." *Review of Economic Studies* 39 (2): 185–192.

Mas-Colell, Andreu, Michael Whinston, and Jerry Green. 1995. *Microeconomic Theory.* Oxford: Oxford University Press.

McKenzie, Lionel. 1954. "On equilibrium in Graham's model of world trade and other competitive systems." *Econometrica* 22 (2): 147–161.

Parente, Stephen L., and Edward C. Prescott. 1994. "Barriers to technology adoption and development." *Journal of Political Economy* 102(2): 298–321.

Prescott, Edward C., and Robert M. Townsend. 1984. "General Competitive Analysis in an economy with private information." *International Economic Review* 25(1): 1–20.

Robinson, Joan. 1933. *The Economics of Imperfect Competition.* London: Macmillan.

Rosen, Sherwin. 1974. "Hedonic prices and implicit markets: Product differentiation in pure competition." *Journal of Political Economy* 82(1): 34–55.

Samuelson, Paul A. 1947. *Foundations of Economic Analysis.* Cambridge, MA: Harvard University Press.

Scarf, Herbert E., and Terje Hansen. 1973. *The Computation of Economic Equilibria.* New Haven: Yale University Press.

Schmeidler, David. 1969. "Competitive equilibria in markets with a continuum of traders and incomplete preferences." *Econometrica* 37: 578–585.

Sonnenschein, Hugo. 1971. "Demand theory without transitive preferences, with applications to the Theory of Competitive Equilibrium." Chapter 10 in *Preferences, Utility, and Demand,* ed. J. S. Chipman et al. New York: Harcourt-Brace-Jovanovich.

———. 1972. "Market Excess demand functions." *Econometrica* 40: 549–563.

Vind, Karl. 1964. "Edgeworth-Allocations in an exchange economy with many traders." *International Economic Review* 5(2): 165–77.

von Neumann, John, and Oskar Morgenstern. 1944. *Theory of Games and economic behavior.* Princeton: Princeton University Press.

Editorial Note

We have rectified a number of misspelled words in the original journal publications, keeping the changes minimal. Three corrections are noted in the text.

— 1 —

An Equilibrium Existence Theorem without Complete or Transitive Preferences*

The Walrasian equilibrium existence theorem is reproved without the assumptions of complete or transitive preferences.

1. Introduction

The aim of this paper is described in its title—to demonstrate that for the general equilibrium Walrasian model to be well defined and consistent (i.e., for it to have a solution), the hypotheses of completeness and transitivity of consumers' preferences are not needed.

Schmeidler (1969) proved the existence of equilibria in a model with a continuum of traders and incomplete preferences; he asked if completeness could be similarly dropped in the case of a finite number of traders; the result here answers this question in the affirmative. We refer to Aumann (1964, also 1962), for forceful arguments in favor of relaxing completeness assumptions on decision-makers' preferences.

For the case where preferences are complete, Sonnenschein (1971) showed how it was possible to obtain continuous demand functions without making any use of transitivity.

Journal of Mathematical Economics 1 (1974), 237–246.

* The content of this paper has been presented at seminars in Berkeley and Stanford; I am indebted to its participants for many helpful comments. I want to thank R. Aumann, G. Debreu, D. Gale, R. Mantel, and B. Peleg for their suggestions; in particular, D. Gale pointed out an error in an earlier version, and R. Aumann and B. Peleg suggested a simplification of the proof. Needless to say, they are innocent of any remaining shortcomings. Research has been supported by NSF Grants GS-40786X and GS-35890X which is gratefully acknowledged.

Let Ω be the consumption set. Our hypotheses on preferences are: $\succ$ is an open subset of $\Omega \times \Omega$ (continuity), which is irreflexive (i.e., $x \succ x$ never holds) and such that, for every $x \in \Omega$, $\{y \in \Omega : y \succ x\}$ is non-empty and convex. We do not assume that $\succ$ is asymmetric or transitive and the stringent convexity hypothesis '$\{y \in \Omega : (x, y) \notin \succ\}$ is convex' is not made; this last condition lacks intuitive appeal in a context where preferences may not be complete. In a few words, the only things of substance we are postulating are non-saturation and the convexity of 'preferred than' sets.

The problem at hand seems to require an existence proof of a novel type. Even assuming transitivity and monotonicity of preferences, their incompleteness may severely destroy the convex-valuedness of the demand correspondence [which is an irrelevant consideration in the continuum of traders context; this is the fact exploited by Schmeidler (1969)]; actually, we shall argue in the appendix, by an example, that an attempt to a proof through demand correspondence is completely barren. Of course, the demonstration we give is a fixed-point one, but the mapping constructed does not appear to have been used before. Perhaps the closest relative to the approach taken here, is Smale's (1974, appendix) existence proof; the specifics are very different, but there is some analogy in the nature of the problems being solved. This will become clear in the text.

For the sake of clarity and conciseness the analysis is limited to pure exchange economies. There is no difficulty in extending the results to, for example, the private ownership economies of Debreu's *Theory of value* (1959).

2. The Model and Statement of Theorems

There are ℓ commodities, indexed by h, and N consumers, indexed by i; $\Omega = R^l_+$.[1]

In section 2.1, a model where consumers are described by preference

1. Commodities will be denoted by superscripts while subscripts will be reserved for (consumption, production) vectors; $x \gg y$ means $x^h > y^h$ for all h, $x > y$ means $x^h > y^h$ for all h and $x \neq y$, $x \geq y$ means $x > y$ or $x = y$; co D, Int D, ∂D stand for the convex hull, the interior, and the boundary of $D \subset R^n$, respectively. The Euclidean norm is $\|\ \|$; for $x,y \in R^n$, xy denotes the inner product. If $B,D \subset R^n$, $B + D = \{z_1 + z_2 : z_1 \in B, z_2 \in D\}$, $BD = \{zy : z \in B, y \in D\}$. When there is no ambiguity, we write b instead of $\{b\}$. If $B \subset R^n$ and $s \in R^n$, $B \gg s$ means $b \gg s$ for every $b \in B$; analogously for $B \geq s$; [ ], [ ), . . . denote segments in the usual way.

relations is given and a theorem is stated. In section 2.2, an alternative model, differing only in the specification of consumers, is described and another theorem stated; it is shown then that the last implies the former.

2.1

Every consumer i is specified by $(X_1, \succ_i, \omega_i)$, where $X_i \subset R^l$, $\succ_i \subset X_i \times X_i$, and $\omega_i \in R^l$. We denote $(z, v) \in \succ_i$ by $z \succ_i v$.

(C.1) *For every i, X_i (the consumption set) is a non-empty, bounded below, closed, convex set.*
(C.2) *For every i, $\succ_i$ (the preference relation) is a relatively open set such that, for every $x_i \in X_i$, $\{z \in X_i : z \succ_i x_i\}$ is non-empty, convex (non-saturation and convexity) and does not contain x_i (irreflexivity).*
(C.3) *For every i, $\omega_i \gg x_i$ for some $x_i \in X_i$.*

An economy $\mathscr{E}$ is identified with $\{(X_i, \succ_i, \omega_i)\}_{i=1}^N$. Let $\Delta = \{p \in \Omega : \sum_{h=1}^{i} p^h = 1.\}$

Definition. $(x, p) \in \prod_{i=1}^{N} X_i \times \Delta$ *is an equilibrium for* $\mathscr{E} = \{(X_i, \succ_i, \omega_i)\}_{i=1}^N$ *if:*

(E.1) $\sum_{i=1}^{N} x_i \leq \sum_{i=1}^{N} \omega_i$;
(E.2) *for every i,* $px_i = p\omega_i$;
(E.3) *for every i, if* $z \succ_i x_i$*, then* $pz > px_i$.

Theorem 1. If $\mathscr{E} = \{(X_i, \succ_i, \omega_i)\}_{i=1}^N$ *satisfies (C.1), (C.2), (C.3), (C.4), then there is an equilibrium for* $\mathscr{E}$.

Let $S = \{x \in R^l : \|x\| = 1\}$ and $\mathscr{E}$ be an economy satisfying the conditions of the theorem. For every i define $g_i : X_i \to S$ by $g_i(x_i) = \{p \in S :$ if $z \succ_i x_i$, then $pz \geq px_i\}$; then equilibrium condition (E.3), in the presence of (C.3), amounts to requiring $(1/\|p\|)p \in g_i(x_i)$. This heuristic comment motivates the more general model of the next section.

2.2

A set $H \subset R^n$ is contractible if the identity map on H is homotopic to a constant map. For the present purposes it will suffice to know that convex sets and intersections of S with convex cones which are not linear subspaces, are contractible. A correspondence is said contractible-valued if its values are contractible sets. The product of two contractible sets is contractible.

For every i let $g_i : X_i \twoheadrightarrow S$ be a correspondence and, in the definitions of section 2.1, substitute $\succ_i$ throughout by g_i.

Replace (C.2) by

(C.2′) *for every i*, $g_i : X_i \twoheadrightarrow S$ *is an u.h.c., contractible-valued correspondence such that, for every* $x_i \in X_i$, *the (possibly empty) set* $\{p \in S : pz \geq px_i$ *for all* $z \in X_i\}$ *is a subset of* $g_i(x_i)$;

and (E.3) by

(E.3′) for every i, $(1/\|p\|)p \in g_i(x_i)$.

We can state:

Theorem 2. *If* $\mathscr{E} = \{(X_i, g_i, \omega_i)\}_{i=1}^N$ *satisfies (C.1), (C.2′), (C.3), (C.4), then there is an equilibrium for* $\mathscr{E}$.

Theorem 2 implies Theorem 1. Let an economy $\mathscr{E} = \{(X_i, \succ_i, \omega_i)\}_{i=1}^N$ satisfying (C.1), (C.2), (C.3), (C.4) be given. For every i and $x_i \in X_i$ define $g_i(x_i) = \{p \in S : \text{if } z \succ_i x_i, \text{ then } pz \geq px_i\}$; since $\succ_i$ is irreflexive and $\{z \in X_i : z \succ_i x_i\}$ is a non-empty, open, convex subset of the convex set X_i, the set $\{p \in R^l : \text{if } z \succ_i x_i, \text{ then } pz \geq px_i\}$ is a non-empty, convex cone which cannot be a linear subspace; therefore $g_i(x_i)$ is non-empty and contractible. Obviously, $\{p \in S : pz \geq px_i \text{ for all } z \in X_i\} \subset g_i(x_i)$.

The correspondence $g_i : X_i \twoheadrightarrow S$ *so defined is u.h.c.*

Proof. The set $\{(z, x_i, p) \in X_i \times X_i \times S : z \succ_i x_i, pz < px_i\}$ is open. The graph of g_i is the complement of the projection of this set on the last two coordinates; hence it is closed. Q.E.D.

Therefore $\mathscr{E}' = \{(X_i, g_i, \omega_i)\}_{i=1}^N$ satisfies the hypothesis of Theorem 2, and so there is an equilibrium for $\mathscr{E}'$.

An equilibrium for $\mathscr{E}'$, *is an equilibrium for* $\mathscr{E}$.

Proof. Let (x, y, p) be an equilibrium for $\mathscr{E}'$. It has to be shown that, for every i, if $z \succ_i x_i$, then $pz > px_i$. Let $pz \leq px_i$, $z \succ_i x_i$, for some i. Pick $\bar{z} \ll \omega_i$, $\bar{z} \in X_i$; then (since $p\bar{z} < p\omega_i \leq px_i$) for every $z' \in [\bar{z}, z]$ we have

"$z' \in X_i$ and $pz' < px_i$" and, for $z' \in [\bar{z}, z]$ sufficiently close to z, $z' \succ_i x_i$; hence $p \notin g_i(x_i)$. A contradiction. Q.E.D.

It is worth emphasizing that the model with consumers specified by the g_i's correspondences admits of more interpretations than the one of (global) preference maximization. For example, it encompasses Smale's notion of an 'extended equilibria' for non-convex economies [in which case g_i would be a normalized gradient vector field; see Smale (1974)] or the various concepts of 'local preference satisfaction' and 'preference fields' to be found in the non-integrability literature [see Georgescu-Roegen (1936); Katzner (1970, ch. 6)]. See also Debreu (1972) from where the notation g_i is taken.

3. Proof of the Theorems

It suffices to prove Theorem 2.

Let $\mathscr{E} = \{(X_i, g_i, \omega_i)\}_{i=1}^N$ satisfy (C.1), (C.2′), (C.3), and (C.4). Denote $e = (1, \ldots, 1) \in R^l$. For every i, define $\bar{g}_i : \Omega \to S$ by

$$\bar{g}_i(x_i,= \begin{cases} g_i(x_i), & \text{if } x_i \in X_i; \\ \{p \in S : pz \geq px_i \text{ for all } z \in X_i\}, & \text{if } x_i \notin X_i, x_i \geq se; \\ \left\{\left(1/\|p\|\right)p : p \in \Delta \text{ and } px_i = \min_h x_i^h\right\}, & \text{if } \min_h x_i^h < s. \end{cases}$$

The correspondence $\bar{g}_i$ satisfies (C.2′) with respect to Ω. Moreover, if $x_i \in \partial\Omega$, then $\bar{g}_i(x_i) = \{p \in S : px_i = 0\}$.

If $(x, p) \in \Omega^N \times \Delta$ is an equilibrium for $\mathscr{E}' = \{(\Omega, \bar{g}_i, \omega_i)\}_{i=1}^N$, then it is an equilibrium for $\mathscr{E}$.

Proof. For every i, $px_i \geq p\omega_i > \operatorname{Inf} pX_i$. Let $x_i \notin X_i$; since $p \in \bar{g}_i(x_i)$, it follows by the construction of $\bar{g}_i$ that if $z \in X_i$, then $pz \geq px_i$, i.e., $px_i \leq \operatorname{Inf} pX_i$; a contradiction. Therefore $x_i \in X_i$, $p \in g_i(x_i)$, for every i. Q.E.D.

In view of this we assume for the rest of the proof that $\mathscr{E}$ satisfies:

(C.5) *For every i, $X_i = \Omega$ and, for all $x_i \in \delta\Omega$, $g_i(x_i)x_i \leq 0$.*

The fixed-point theorem to be used (an immediate corollary of the Eilenberg–Montgomery theorem) is contained in the:

Lemma. *If $K \subset R^n$ is a non-empty, convex, compact set and $F : K \to R^n$ is an u.h.c. contractible-valued correspondence, then there exist $x \in K$ and $y \in F(x)$ such that $(y - x)(z - x) \leq 0$ for all $z \in K$.*

Proof. Let $F(K) \subset H$, where H is taken convex and compact. Since K is closed and convex, we can define a continuous function $\sigma_K : H \to K$ by letting $\sigma_K(x) \in K$ be the $\| \; \|$-nearest element to x in K. For every $x \in H$, $z \in K$, one has $(x - \sigma_K(x))(z - \sigma_K(x)) \leq 0$. The correspondence $F \circ \sigma_K : H \to H$ is u.h.c. and contractible-valued. By the Eilenberg–Montgomery fixed-point theorem [Eilenberg (1946); see also, Debreu (1952)], there is $x \in H$ such that $x \in F(\sigma_K(x))$.

Call $y = \sigma_K(x)$. Then $x \in F(y)$, $y \in K$ and $(x - y)(z - y) \leq 0$ for every $z \in K$. Q.E.D.[2]

Pick an arbitrary $\varepsilon > 0$ and let $re > \sum_{i=1}^{N} \omega_i + \varepsilon e$, $k = r\ell(\ell N + 2)$ and $\varphi : \Delta \to R^l$ be a continuous function such that: (i) $p\varphi(p) \geq 0$ for all $p \in \Delta$; (ii) $\varphi(\Delta) \leq \varepsilon e$; (iii) if $p^h = 0$ then $\varphi^h(p) \leq -k$, for all $p \in \Delta$. It is clear that such a function exists.

By an obvious limiting argument the theorem will be proved if we show the existence of $(x, p) \in \Omega^N \times \Delta$ such that $\sum_{i=1}^{N} x_i \leq \sum_{i=1}^{N} \omega_i + \varepsilon e$ and $p \in g_i(x_i)$, $|px_i - p\omega_i| < \varepsilon$ for all i.

Denote $\alpha_i(p) = p\omega_i + (1/N)p\varphi(p)$ and define the u.h.c., contractible-valued correspondences $\Phi_i : [0, r]^l \times \Delta \to R^l$, $1 \leq i \leq N$, $\Phi_\Delta : [0, r]^{lN} \times \Delta \to R^l$ by

$$\Phi_i(x_i, p) = x_i + \alpha_i(p) g_i(x_i) - px_i \frac{p}{\|p\|},$$

$$\Phi_\Delta(x, p) = p + \sum_{i=1}^{N} (x_i - \omega_i) - \varphi(p).$$

Let then the u.h.c., contractible-valued correspondence $\Phi : [0, r]^{lN} \times \Delta \to R^{l(N+1)}$ be given by

2. The lemma can be proved by appealing only to the Brouwer fixed-point theorem. Suppose F is a function. Then there is $x \in K$ such that $x = \sigma_K(F(x))$, i.e., $(F(x) - x)(z - x) \leq 0$ for all $z \in K$; the result follows then by a continuity argument and the fact [proved, for example, in Mas-Colell (1974b)] that, with the hypothesis made, there is for any $\varepsilon > 0$ a continuous function $f : K \to R^n$ which graph is contained in the ε-neighborhood of the graph of F.

$$\Phi(x, p) = \Phi_1(x_1, p) \times \ldots \times \Phi_N(x_N, p) \times \varphi_\Delta(x, p).$$

Applying the lemma to Φ we obtain $p \in \Delta$, $x \in \Omega^N$, and, for every i, $p_i^* \in g(x_i)$ such that, denoting $\hat{x} = \sum_{i=1}^N x_i$, $\hat{\omega} = \sum_{i=1}^N \omega_i$,

(a) for every i, if $z \in [0, r]^l$, then $(z - x_i)\left(\alpha_i(p)\, p_i^* - px_i \frac{p}{\|p\|}\right) \leq 0$;

(b) if $q \in \Delta$, then $(q - p)(\hat{x} - \hat{\omega} - \varphi(p)) \leq 0$.

We show that (x, p) is as desired; let $y = \hat{\omega} + \varphi(p)$.

If $p \in \partial\Delta$, then

$$\left(\frac{1}{\ell}e - p\right)(\hat{x} - y) \geq -p\hat{x} - \frac{1}{\ell}ey \geq -\ell Nr - r + \frac{k}{\ell} = \frac{r}{\ell} > 0,$$

which contradicts (b). Therefore $p \geq 0$ and, by (b), $\hat{x} - y = \lambda e$ for some $\lambda \in R$.

If $x_i \in \partial\Omega$, then by taking $z = 0$ (resp. $z = re$) if $x_i \neq 0$ (resp. $x_i = 0$), we contradict (a). Therefore, for every i, $x_i \gg 0$ and so, by (a), $\alpha_i(p)p_i^* \geq (px_i/\|p\|)p$. Hence $\alpha_i(p) \geq px_i$ for every i, which implies $\lambda \leq 0$. From this we get, for every i, $x_i \leq \hat{x} \leq y < re$, and so, again by (a), $\alpha_i(p)p_i^* \leq (px_i/\|p\|)p$. Therefore, for every i, $\alpha_i(p)p_i^* = (px_i/\|p\|)p$ which yields $\alpha_i(p) = px_i$, $p = p_i^*$. Since, then, $p\hat{x} = py$, we also have $\lambda = 0$. Hence $\sum_{i=1}^N x_i \leq \sum_{i=1}^N \omega_i + \varepsilon e$ and $p \in g_i(x_i)$, $p\omega_i \leq px_i \leq p\omega_i + \varepsilon$ for all i. This concludes the proof.

4. Remarks

4.1

Let $\succ$ be a preference relation on Ω (to make things specific) satisfying the hypotheses of Theorem 1, i.e., (C.2). For $p \in \Delta$, $p \gg 0$, $w \in [0, \infty)$ define $h^{\succ}(p, w) = \{x \in \Omega : \text{if } y \succ x, \text{ then } py > px\}$; this set is non-empty [this follows from Sonnenschein's proof in (1971); although his result is phrased in terms of a complete preorder $\succsim$, the proof of the non emptiness of $h^{\succ}(p, w)$ uses only the convexity of the induced $\succ$]. Hence a demand correspondence $h^{\succ}$: Int $\Delta \times [0, \infty) \to \Omega$ is well defined; it is also u.h.c. Given initial endowments $\omega \in \Omega$, $h^{\succ,\omega}$: Int $\Delta \to \Omega$ stands for the excess demand correspondence generated by $h^{\succ}$, i.e., $h^{\succ,\omega}(p) = h(p, p\omega)$.

In the appendix we give an example of a quite simple $\succ$ which is transitive, monotone and satisfies (C.2), but for some $\omega \in \Omega$, no u.h.c. subcorrespondence of $h^{\succ,\omega}$ is connected-valued (say, that a correspondence is connected-valued if its values are connected sets; $F : A \to B$ is a subcorrespondence of $G : A \to B$ if $F(t) \subset G(t)$ for all $t \in A$); also, $\succ$ possesses a continuous utility function [i.e., a function $u : \Omega \to R$ such that if $x \succ y$, then $u(x) > u(y)$; this is Aumann's term (1964)], but not a quasi-concave one. Thus, the example shows that, even if transitivity is assumed, Theorem 1 cannot be obtained as a corollary of available existence results and, also, that a proof by the way of demand correspondences is not possible.

4.2

Schmeidler (1969) proved that, in the continuum of agents case, equilibria exist for economies which (in addition to other hypotheses) have consumers with transitive, incomplete, not necessarily convex, preferences. Trivial examples [every consumer has preferences on R^2_+ given by $\{(x, y) : \text{'}x^1 > y^1 - 1 \text{ and } x^2 > y^2\text{' or '}x^1 > y^1 \text{ and } x^2 > y^2 - 1\text{'}\}$] show that, unless convexity of preferences is assumed, this result cannot be improved upon by dropping transitivity. This is, we believe, a very good reason to keep transitivity (of $\succ$) among the standard assumptions of equilibrium analysis.

Appendix

We give here an example of a relation $\succ$ on Ω such that (i) $\succ$ is transitive, monotone, and satisfies (C.2); (ii) there is a continuous utility for $\succ$; (iii) there is no quasi-concave utility for $\succ$; and (iv) for a $\omega \in \Omega$, there is no u.h.c. connected-valued subcorrespondence of $h^{\succ,\omega}$.

We define $\succ$ first in R^2_+ and show then that there is no u.h.c., connected-valued subcorrespondence of the *demand* correspondence $h^{\succ}$. This suffices since defining a $\succ'$ in R^3_+ by '$(x^1, x^2, x^3) \succ' (y^1, y^2, y^3)$ if and only if $(x^1, x^2) \succ (y^1, y^2)$' and taking $\omega = (0, 0, 1)$ what was true of the demand correspondence of $\succ$ will be true of the excess demand correspondence of $\succ'$.

Hence, let $\ell = 2$. Define (utility functions) $u', u'' : \Omega \to R$ by $u'(x) = \min\{x^1, 2(x^1 + x^2) - 2\}$, $u''(x) = \min\{x^2, 2(x^1 + x^2) - 2\}$; see fig. 1.

Define $\succ \subset \Omega \times \Omega$ by (see fig. 2) $\succ = \{(x, y) \in \Omega \times \Omega : \text{'}x \gg y\text{'}$ or

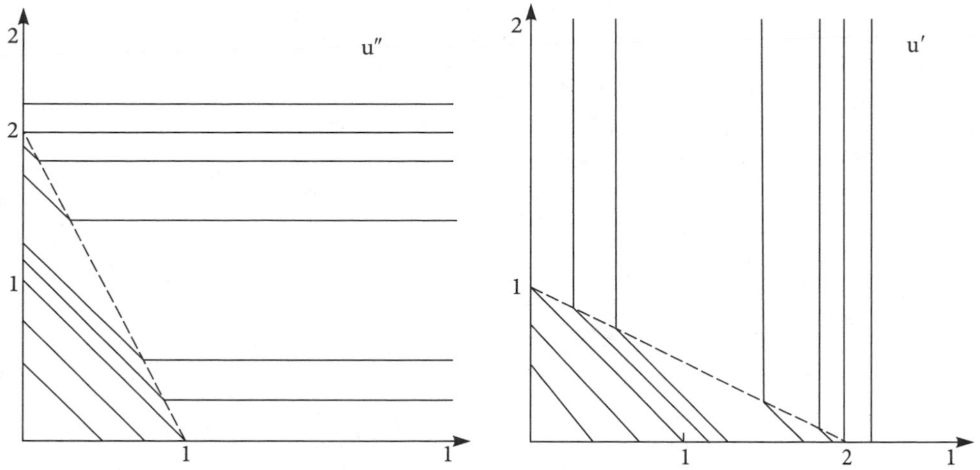

Figure 1

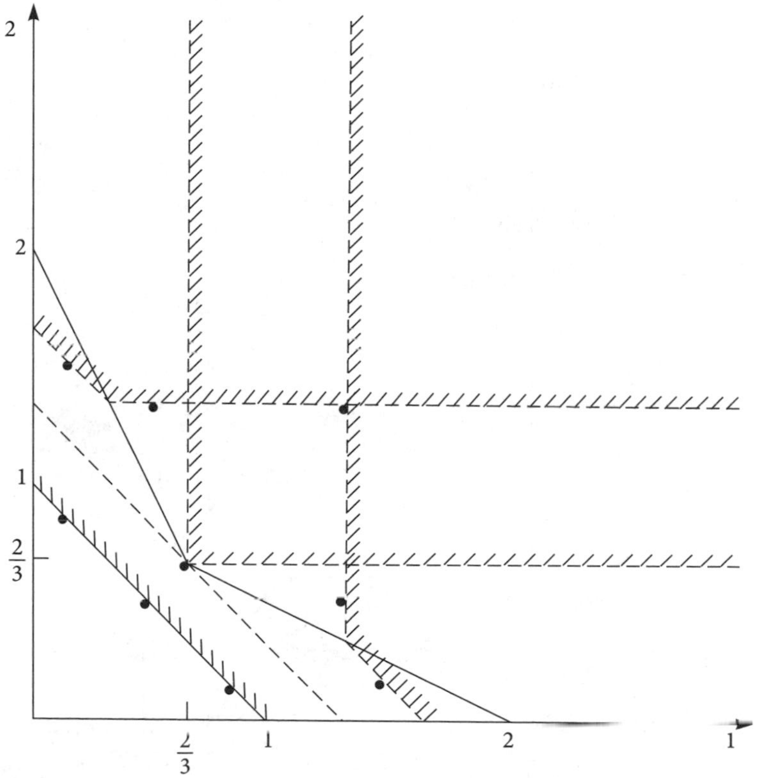

Figure 2

'$u'(x) > u'(y)$ and $y^1 > \frac{2}{3}, y^2 < \frac{2}{3}$' or '$u''(x) > u''(y)$ and $y^2 > \frac{2}{3}, y^1 < \frac{2}{3}$' or '$y^1 + y^2 < \frac{4}{3}$, and $x^1 + x^2 > y^1 + y^2$'}. In the definition of $\succ$ only strict inequalities enter, and so the relation is open; clearly, it is also monotone and irreflexive. It is easily checked that:

(a) If $y^1 + y^2 < \frac{4}{3}$ and $x \succ y$, then $x^1 + x^2 > y^1 + y^2$;

(b) if $y^1 + y^2 \geq \frac{4}{3}, y^1 < \frac{2}{3}$ (resp. $y^2 < \frac{2}{3}$), then $x \succ y$ implies $x^2 > \frac{2}{3}$ (resp. $x^1 > \frac{2}{3}$) and $x^1 + x^2 > \frac{4}{3}$.

The fact that $\succ$ is convex is a consequence of (a) and the monotonicity of u', u''. To see that $\succ$ is transitive, let $z \succ y$, $y \succ x$. If $x \gg \left(\frac{2}{3}, \frac{2}{3}\right)$, then obviously $z \succ x$. If $x^1 + x^2 \geq \frac{2}{3}$ and $x^1 < \frac{2}{3}$, then $y \succ x$ means $u''(x) > u''(y)$ and, by (b), $y^2 > \frac{2}{3}$; therefore $z \succ y$ implies $u''(z) > u''(y)$ and we have $z \succ x$. If $x^1 + x^2 \geq \frac{4}{3}$ and $x^2 < \frac{2}{3}$, a symmetric argument applies. If $x^1 + x^2 < \frac{4}{3}$ and

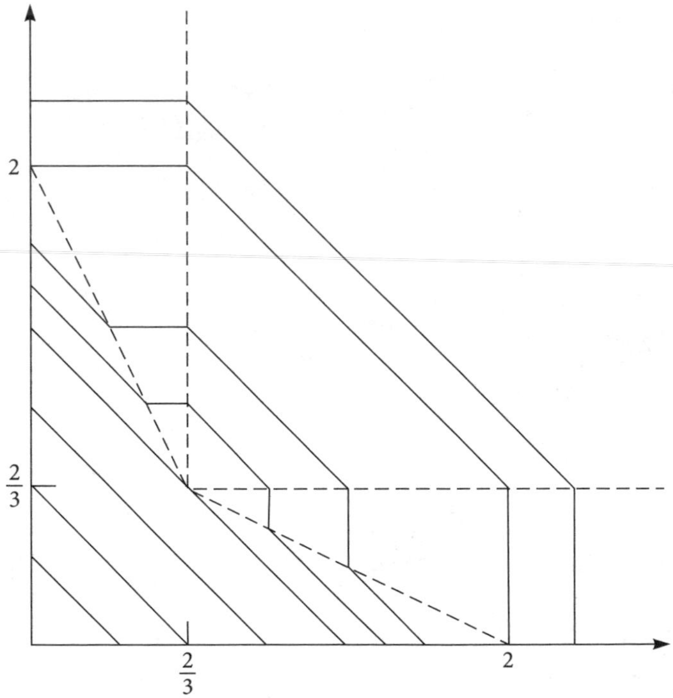

Figure 3

$y^1 + y^2 < \frac{4}{3}$, (then $z \succ x$ by (a) and the definition. If $x^1 + x^2 < \frac{4}{3}$ and $y^1 + y^2 \geq \frac{4}{3}$, then, by (b), $z^1 + z^2 > \frac{4}{3}$, i.e., $z \succ x$.

Take $\bar{p} = \left(\frac{1}{2}, \frac{1}{2}\right)$, $\bar{w} = 1$ and suppose that η : Int $\Delta \times [0, \infty) \to \Omega$ is an u.h.c. connected-valued subcorrespondence of $h^{\succ}$. For $t \in [0, 1)$ take $p(t) = ((t + 1)/2, (1 - t)/2)$, $w(t) = 1 - t$. For every $x \in \Omega$ if $x^1 < \frac{2}{3}$ and $x^2 < 2$, then $(0, 2) \succ (x^1, x^2)$. Therefore, for every $t \in [0, 1)$, if $x \in \Omega$ and $0 < x^1 < \frac{2}{3}$, then $x \notin h^{\succ}(p(t), w(t))$ and for $\bar{t}$ close enough to 1, $h^{\succ}(p(\bar{t}), w(\bar{t})) = \{(0, 2)\}$. Hence $\eta(p(t), w(t)) = \{(0, 2)\}$ for every $t \in [0, 1)$, and so $(0, 2) \in \eta(\bar{p}, \bar{w})$. By a symmetric argument $(2, 0) \in \eta(\bar{p}, \bar{w})$ and therefore $[(0, 2), (2, 0)] \subset \eta(\bar{p}, \bar{w})$. But this is impossible since, for example, $\left(\frac{1}{2}, \frac{3}{2}\right) \notin h^{\succ}(\bar{p}, \bar{w})$. Hence no such η exists.

The relation $\succ$ satisfies Peleg's (1970) spaciousness condition for the existence of a continuous utility function; one is represented in fig. 3. It is immediate that the horizontal and vertical segments of fig. 3 should appear in any indifference map of a utility for $\succ$, hence no quasi-concave utility exists [examples of open relations having quasi-concave but no continuous utility have been given by Schmeidler (1969) and Peleg (1970)].

References

Aumann, R., 1962, Utility theory without the completeness axiom, Econometrica 30, 445–462.

Aumann, R., 1964, Subjective programming, in: M. W. Shelly and G. L. Bryan, eds., Human judgments and optimality (John Wiley, New York) ch. 12.

Debreu, G., 1959, Theory of value (John Wiley, New York).

Debreu, G., 1972, Smooth preferences, Econometrica 40, 603–617.

Eilenberg, S. and D. Montgomery, 1946, Fixed-points theorems for multivalued transformations, American Journal of Mathematics 68, 214–222.

Gale, D., 1955, The law of supply and demand, Mathematica Scandinavica 3, 155–169.

Georgescu-Roegen, N., 1936, The pure theory of consumer's behavior, The Quarterly Journal of Economics, 133–170.

Katzner, D., 1970, Static demand theory (Macmillan, London).

Mas-Colell, A., 1974, A note on a theorem of F. Browder, Mathematical Programming 6, 229–233.

Peleg, B., 1970, Utility functions for partially ordered topological spaces, Econometrica 38, 93–96.

Schmeidler, D., 1969, Competitive equilibria in markets with a continuum of traders and incomplete preferences, Econometrica 37, 578–585.

Smale, S., 1974, Global analysis and economics IIA, Extension of a theorem of Debreu, Journal of Mathematical Economics 1, 1–14.
Sonnenschein, H., 1971, Demand theory without transitive preferences, with applications to the theory of competitive equilibrium, in: J. Chipman et al., eds., Preferences, utility, and demand (Harcourt-Brace-Jovanovich, New York) ch. 10.

— 2 —

A Model of Equilibrium with Differentiated Commodities*

1. Introduction

The equilibrium theory associated with the names of Arrow and Debreu [see Debreu (1959), Arrow and Hahn (1971)] contemplates a world with a finite number of homogeneous and perfectly divisible commodities where traders interact (exclusively) through a price system taken by each one of them as given. This perfectly competitive hypothesis has been justified by the 'core equivalence theorem' of Debreu and Scarf (1963) and Aumann (1964) [for a thorough account see Hildenbrand (1974)]: if no trader arrives to the market with a substantial amount, i.e., a 'corner', of any commodity, then unrestricted bargaining (in the cooperative sense of core theory) leads to perfect competition.

In contrast, imperfect competition theory [in either Chamberlin (1956) or Robinson (1933) version] starts with a very different perception of the economic realm; commodities are not homogeneous but subject to differentiation and, consequently, traders enjoy a certain degree of monopoly with respect to the commodities they control. Still, the monopoly power of every single trader is limited by the existence of substitutability relations among commodities; it is a common contention of imperfect competition

Journal of Mathematical Economics 2 (1975), 263–295.

* Presented at the Mathematical Social Science Board Colloquium on Mathematical Economics in August 1974 at the University of California, Berkeley. The author is indebted to F. Delbaen, B. Grodal, J. Ostroy and H. Sonnenschein for very useful conversations; F. Delbaen, in particular, was very helpful at one important step. Thanks are also due to the audience of a seminar at UCLA and to Professor L. Hurwicz, who kindly allowed me to see some unpublished manuscripts of his. Final responsibility remains with me. Support from NSF grants SOC73-05650A01 and SQC72-05551A02 is gratefully acknowledged.

theory that in a large economy with a large number of mutually substitutive commodities and no 'big' trader, every commodity will be substitutable in the market with infinite elasticity and a perfectly competitive outcome will prevail [see, for an account, Samuelson (1969, p. 135)].

The analogy between the statements of the two previous paragraphs is clear. It is the purpose of this paper to describe a general equilibrium (pure exchange) model exhibiting commodity differentiation as an essential feature and where, if certain crucial assumptions are satisfied, the core is no larger than the set of competitive equilibria.

In a few words, the model is as follows. There are two kinds of commodities, differentiated and homogeneous; it is important that there exists at least one homogeneous commodity and, for specificity, we shall assume there is just one. Differentiated commodities are only available in integral amounts (i.e., they are not perfectly divisible) and they are described by specifying its *characteristics,* a point in an a priori given *compact* metric space of characteristics K. There are infinitely many traders, each one of them initially possessing some amount of commodities; we assume that, in a certain sense, every commodity is available in the market. Traders have preferences continuous on characteristics; so, commodities with close characteristics are close substitutes. A price system is a function continuous also on characteristics. We assume that traders are not too dissimilar; more technically, endowments and preferences belong to a 'compact' set of traders' characteristics [this is a familiar hypothesis, see Hildenbrand (1974), although not in this context, but see Bewley (1970)]. With this set-up, and some further assumptions, we prove the equality of the core and the set of equilibria; we also show that those sets are non-empty, i.e., the result is not vacuous. To make the existence of equilibria problem a tractable one, we exploit the concept of equilibrium distributions recently introduced by Hart, Hildenbrand and Kohlberg (1974).

To be sure, the idea of a space of commodity characteristics is not new. In one form or another it appears whenever quality differentials (or spatial matters) are focused upon; it can be found in Houthakker (1952), Lancaster (1966), or in the more recent literature on hedonic prices, see Rosen (1974) and his references. Usually the space of characteristics will be a subset of $\mathbf{R}^n$; for our purposes, however, and with no further complication of the proof, it suffices that it be a metric space.

From the mathematical point of view this paper amounts to yet another extension of Aumann's equivalence theorem (plus existence) to an infinite-dimensional commodity space context. Indeed, we find it convenient to treat individual commodity bundles as (purely atomic) measures on the space of characteristics, i.e., an individual commodity bundle specifies for each set of characteristics the total amount of commodities with characteristics in this set. The (individual) consumption set becomes then a subset of the linear space of measures on K (which we endow with the weak-star topology). In different directions infinite-dimensional versions of Aumann's theorem have been given by Bewley (1973), Gabszewicz (1968) and Mertens; their models, however, have time or uncertainty in mind; it is appropriate, then, as Bewley and Gabszewicz argue, to take some $\mathscr{L}^{\infty}(M, \mathscr{M}, \mu)$ space as a commodity space and to assume that initial endowments are strictly positive functions. Clearly, such a model is inadequate for the differentiated commodities situation; the choice problem in the latter is not typically how much of each (perhaps individually insignificant) commodity to buy, but which commodity to buy. An existence theorem for an infinite-dimensional case has been given by Bewley (1972).

Ostroy (1973) has provided examples where, in a model addressing a situation analogous to ours, the core is much larger than the set of equilibria; what happens, in essence, is that every trader is endowed with commodities that, from the point of view of trader tastes, are complementary to any other commodity so that the possibility of competitiveness among them is very limited. Ostroy's examples underline the crucial role of the *compactness* assumptions in our model; since the space of commodity characteristics is compact, we have that, although infinite-dimensional, the 'size' of the commodity space is sufficiently small relative to the size of the economy for the equality of core and equilibria to obtain [this is a well-known heuristic requirement for Aumann's theorem to be generalizable; see Bewley (1973), Gabszewicz (1968)]; also, since the set of agents' characteristics is compact, the relations of substitutability among commodities are unambiguous: if the characteristics of two commodities are sufficiently close there is no group of traders which cannot easily substitute one for the other; an example in the appendix will emphasize the importance of this fact.

A further word on the compactness hypotheses. They clearly amount to the a priori imposition of a certain uniformity on people's tastes. In this respect, our model is less general than the Arrow–Debreu theory, where the relations of substitutability are treated implicitly and with full generality (two commodities are substitutes if and only if traders treat them as such). Our uniformity condition is probably a reasonable price to pay in order to be able to treat substitutability explicitly and to say something interesting about, say, continuity of price equilibria with respect to characteristics.

The indivisibility assumption on differentiated commodities may appear odd and, indeed, to the extent that our model includes the non-differentiated one, it is a complicating factor [see Henry (1970), Dierker (1971), Broome (1972)]. It is, however, of decisive help with the commodity differentiation aspect of the model; it simplifies matters, not only technically, but also conceptually. Since, as a practical matter, any instance where product differentiation becomes interesting involves indivisibilities, the assumption is probably acceptable. We believe that to encompass differentiated commodities which are available in 'infinitesimal' amounts a substantial rethinking of the model is needed.

A number of further restrictive assumptions will be made. They will be spelled out in the main body of the text. Although technically important and even crucial, they are not related to the essence of the problem.

Section 2 describes the model and section 3 contains the proofs; an example is given in the appendix. The strategy of the proofs is plain enough: the commodity space is approximated by a finite-dimensional one and the validity of the theorems is established by a suitable pass to the limit.

2. The Model

2.1. Commodity Characteristics

(I) *There is given a compact metric space K, called space of commodity characteristics;* d_K *denotes a metric for K.*

A point of K is to be understood as a complete description of a unit of a commodity; two commodities are close if the characteristics of their units are topologically close in K. Without possible confusion we will often refer to points in K as commodities.

We postulate the existence of an homogeneous commodity h not subject to differentiation.

(II) $K = K^d \cup \{h\}$, *where* K^d *is compact.*

Generic elements of K are denoted by s.

2.2. Individual Commodity Bundles

An individual commodity bundle (i.c.b.) assigns a non-negative quantity to each commodity. We shall assume, first, that the total number of units of commodities in an i.c.b. is finite and, second, that commodities distinct from h, i.e., differentiated commodities, enter the i.c.b. only in integral amounts (this embodies a uniform discreteness assumption for differentiated commodities). Thus, for a given i.c.b., there is only a finite number of commodities with non-zero quantities attached.

It is both convenient (this, for the moment, has to be taken on faith) and natural to describe formally an i.c.b. as a measure, i.e., as a rule that assigns to each set of characteristics the total number of units of commodities with characteristics in this set.

Let $\mathscr{B}(K)$ be the Borel σ-field of subsets of K.

(III) *An individual commodity bundle a is a non-negative, bounded (Borel) measure on K such that the restriction to K^d, denoted $a \mid K^d$, is integer valued.*

Hence, an i.c.b. is formed by a finite number of atoms. We write $a(s)$ for $a(\{s\})$ and denote by b the commodity bundle which assigns 1 to h and nothing to K^d, i.e., $b(h) = 1$, $b(K^d) = 0$. Individual commodity bundles are added in the obvious way, i.e.,

$$(a + a')(V) = a(V) + a'(V).$$

2.3. Consumption Set

(IV) *The consumption set, denoted Ω, is the set of all individual commodity bundles a such that $a(K^d) \leq \alpha$, where α is a given positive integer.*

The intended interpretation is that α is a very large number. The restriction to bundles bounded on K^d is not essential but it facilitates the arguments very much and no substantial loss of economic generality is implied.

Let $m(K)$ be the set of bounded, signed (Borel) measures on K; endowed with the variation norm $m(K)$ is the dual of the separable Banach space $C(K)$—the space of continuous functions on K—with the supremum norm (Riesz representation theorem). The weak-star topology of $m(K)$ is the topology of pointwise convergence on the continuous functions, i.e., $\mu_n \to \mu$ if and only if

$$\int f \, d\mu_n \to \int f \, d\mu \quad \text{for every} \quad f \in C(K).$$

The weak-star topology of $m(K)$ is separable and norm-bounded subsets of $m(K)$ are compact and metrizable; the same is true of $m(K^d)$, defined

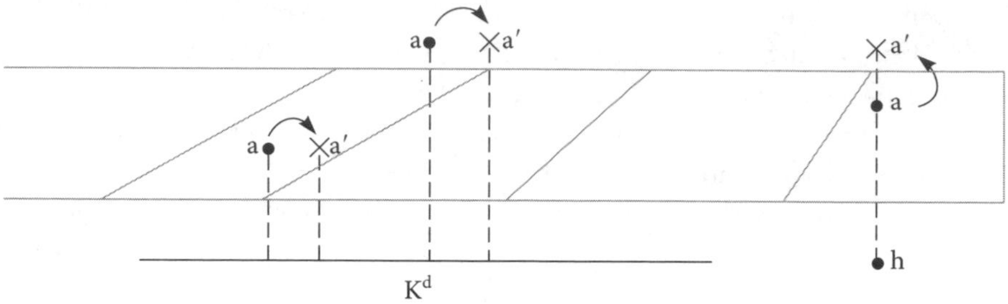

Figure 1

accordingly. [See, for example, Royden (1968, ch. 14) and Schaefer (1970, ch. III, sec. 4).]

The natural notion of 'closeness' for i.c.b. is the following: a' is close to a if $a'(h)$ is close to $a(h)$ and $a' \mid K^d$ is obtained from $a \mid K^d$ *by* perturbing slightly the characteristics of the units of the commodities in $a \mid K^d$ (see fig. 1). This concept of closeness is *exactly* captured by the weak-star topology on Ω.

(V) *Ω, which is a subset of $m(K)$, is endowed with the weak-star topology, i.e., $a_n \to a$ if*

$$\int f\,da_n \to \int f\,da \quad \textit{for every} \quad f \in C(K).$$

Let d'_Ω be a metric for the weak-star topology on the α ball of $m(K^d)$; any two such metrics are equivalent. We let a metric d_Ω on Ω be given by

$$d_\Omega(a, a') = d'_\Omega(a \mid K^d, a' \mid K^d) + \|a(h) - a'(h)\|,$$

where $\| \ \|$ is the Euclidean norm.

2.4. *Preferences*

(VI) *A preference relation is a complete preorder $\succsim \subset \Omega \times \Omega$ such that:*

(i) *$\succsim$ is closed (continuity);*
(ii) *if $a' \geq a$ and $a'(h) > a(h)$, then $a' \succ a$ (i.e., no $a \succsim a'$);*[1]
(iii) *if $a'(h) > 0$ and $a(h) = 0$, then $a' \succ a$;*
(iv) *for any $a \in \Omega$, there is a real $\gamma > 0$ such that $\gamma b \succ a$;*

1. For $x, y \in m(K)$, $x \leq y$ means $x(V) \leq x(V)$for every $V \in \mathscr{B}(K)$.

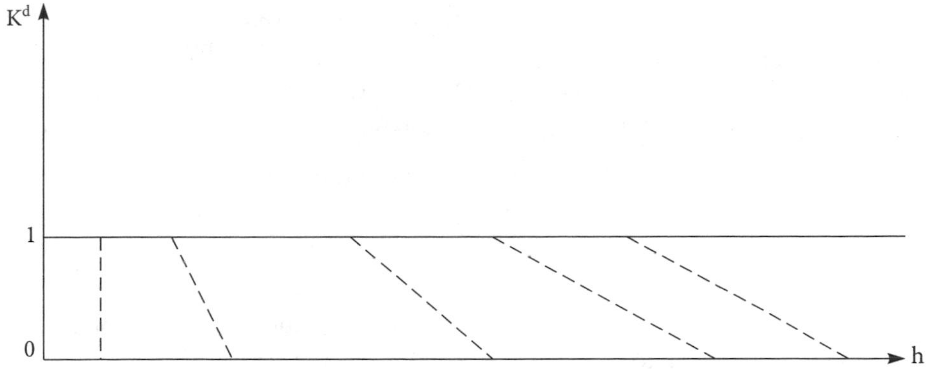

Figure 2

(v) *there is a real* $\xi > 0$ *such that if* $a(h) = a'(h)$ *and* $d'_{\Omega}(a \mid K^d, a' \mid K^d) \leq 1/\xi$, *then* $a + \xi b \succ a'$.

For K^d a singleton, fig. 2 describes the kind of (severely restricted) preferences this paper deals with.

Some comments are called for:

(1) Assumptions (i) and (ii) are standard (once it has been accepted that 'close' for i.c.b. means weak-star close!). Note that we assume desirability only for commodity *h*.

(2) Assumptions (iii) and (iv) are needed even if there is no differentiation (i.e., K^d is discrete); they are related to the indivisibility aspect of the model; this being so we have not tried to develop more acceptable conditions but we are aware that (iii) is particularly inadmissible.

(3) Assumption (v) holds vacuously if K^d is discrete and on bounded subsets of Ω it is implied by (i) and (ii). The condition depends only on the topology of Ω and not on the particular metric chosen. The following condition implies both (iv) and (v) and may be more palatable (see fig. 2):

(vi) there is a real $\xi > 0$ such that $(a(h) + \xi)b \succ a$ for all $a \in \Omega$.

More comments on this assumption will be given in section 2.5.

Denote by $\mathscr{P}$ the set of preference relations.

Let $\mathscr{C}(\Omega \times \Omega)$ be the space of nonempty closed subsets of $\Omega \times \Omega$; with the closed convergence topology, $\mathscr{C}(\Omega \times \Omega)$ becomes a separable, compact metric space [see Hildenbrand (1974, Part I, BII, Thm. 2)]. Since $\mathscr{P} \subset \mathscr{C}(\Omega \times \Omega)$, we can endow $\mathscr{P}$ with the relativized closed convergence topology; let $d_{\succsim}$ be a metric for this topology.

(VII) *$\mathscr{P}$ is endowed with the closed convergence topology, i.e.,*

$$\succsim_n \to \succsim \quad \text{if} \quad d_{\succsim}(\succsim_n, \succsim) \to 0.$$

We refer to Hildenbrand (1974) and Grodal (1974) for an extensive discussion of topological spaces of preferences. The appropriateness of the closed convergence to topologize $\mathscr{P}$ rest on the same arguments (closedness of demand correspondences, etc.) than the ones for the finite and homogeneous commodities case. Some of them will be spelled out in the proofs.

2.5. Consumer Characteristics

(VIII) *There are given two sets $E \subset \Omega$ and $P \subset \mathscr{P}$ satisfying:*

- (i) *E is compact;*
- (ii) *P is compact;*
- (iii) *condition (v) in the definition of preferences holds uniformly on P, i.e., there is $\xi > 0$ such that if $\succsim \in P$, $a(h) = a'(h)$ and $d'_\Omega(a|K^d, a'|K^d) \leq 1/\xi$, then $a + \xi b \succ a'$.*

A consumer is a pair $(\succsim, \omega) \in P \times E$.

Conditions (i) and (ii) are familiar in equilibrium analysis [see Hildenbrand (1974)]; they mean that agents' characteristics are not 'too dissimilar'. Condition (iii) imposes, in a uniform fashion, restrictions of a particular, and somewhat ad hoc, kind on the asymptotic [i.e., for i.c.b. with very large $a(h)$] form of the allowable preference relations. Its role is to permit the confinement of all the arguments in the proofs to a bounded subset of Ω. Unfortunately, while the need for our results of the compactness assumptions (i) and (ii) can be argued by example (see Appendix) or intuitively (see Introduction), the dispensability of (iii) is an open question; for the existence result, it is dispensable in the very special case where $\omega \in E$ implies $\#\omega(K^d) \leq 1$, i.e., when consumers can possess, before exchange, at most one unit of commodities other than h.

Of course, we could replace (iii) by the stronger [see (3) of section 2.4]:
(iv) there is a real $\xi > 0$ such that for every $\succsim \in P$ and $a \in \Omega$, $a(h) + \xi b \succ a$.

2.6. Aggregate Commodity Distribution

The Borel σ-field on Ω generated by the weak-star (relatively) open subsets of Ω is denoted $\mathscr{B}(\Omega)$. Let $i : \Omega \to m(K)$ be the inclusion map.

For every $V \in \mathscr{B}(K)$ the function $a \mapsto a(V)$ from Ω to R is measurable (Fact 1 in section 3.1). Let η be a probability measure on $(\Omega, \mathscr{B}(K))$; if

$$\textstyle\int_\Omega a(h)\, d\eta(a) < \infty,$$

we denote by $\int_\Omega i\, d\eta$ the element of $m(K)$ defined by

$$\left(\textstyle\int_\Omega i\, d\eta\right)(V) = \textstyle\int_\Omega a(V)\, d\eta(a), \qquad V \in \mathscr{B}(K)$$

(Fact 2 in section 3.1). From now on, if we write $\int_\Omega i\, d\eta$, we imply $\int_\Omega a(h)\, d\eta(a) < \infty$.

(IX) *Let η be a probability measure on $(\Omega, \mathscr{B}(\Omega))$, if $\int_\Omega a(h)\, d\eta(a) < \infty$, then $\int_\Omega i\, d\eta$ is called the aggregate commodity distribution corresponding to η.*

2.7. Economy

Endowing $P \times E$ with the product topology, $\mathscr{B}(P \times E)$ is the corresponding Borel σ-field.

(X) *The economy is a probability measure ν on $(P \times E, \mathscr{B}(P \times E))$ such that*

$$\textstyle\int_\Omega a(h)\, d\nu_E(a) < \infty,$$

and

$$\operatorname{supp}\left(\textstyle\int_\Omega i\, d\nu_E\right) = K,$$

where ν_E is the marginal distribution of ν on E.

Note that the conditions imposed on aggregate initial endowments are familiar: we require, first, that the mean amount of commodities be finite and, second, that 'every commodity be available on the market'.

2.8. Prices

Let $C^+(K) = \{p \in C(K) : p \geq 0, p(h) > 0\}$.

(XI) *A price system is an element of* $C^+(K)$.

So, we impose as a condition that similar commodities have similar prices (remember that every commodity is present in the marketplace).

If $x \in m(K)$, $f \in C(K)$, we denote

$$fx = \int_K f \, dx.$$

2.9. Equilibrium Distributions

For $p \in C^+(K)$ let

$$H_p = \{(\succsim, \omega, a) \in P \times E \times \Omega : pa \leq p\omega \text{ and } \text{“}pa' \leq p\omega \text{ implies } a \succsim a'\text{”}\}.$$

We introduce now the concept of 'equilibrium distribution' due to Hart, Hildenbrand and Kohlberg (1974).

(XI) *A probability measure* τ *on* $(P \times E \times \Omega, \mathscr{B}(P \times E \times \Omega))$ *is an equilibrium distribution for the economy* ν *if there is* $p \in C^+(K)$ *such that:*

(i) $\tau_{P \times E} = \nu$;
(ii) $\int_\Omega i \, d\tau_\Omega \leq \int_\Omega i \, d\nu_E$;
(iii) $\tau(H_p) = 1$.

p is then called an equilibrium price system.

2.10. Existence Theorem

Theorem 1. For any economy ν *there is an equilibrium distribution* τ.

2.11. Representation of Economies

Let $I = [0, 1]$ and λ denote Lebesgue measure.

(XII) *A representation for the economy* ν *is a measurable function* $e : I \to P \times E$ *such that* $\nu = \lambda \circ e^{-1}$.

Since $P \times E$ is a complete metric space, any economy ν has a representation [see Hildenbrand (1974, p. 50)].

We usually write $(\succsim_t, \omega(t))$ for $e(t)$.

It is not difficult to check that the measurability condition for e can be expressed in the following more transparent form: e is measurable if and only if (i) for every $B \in \mathscr{B}(K)$, $t \mapsto \omega(t)(B)$ is measurable, and (ii) for every $a, a' \in \Omega$, $\{t : a \succsim_t a'\}$ is a measurable set [those are the measurability conditions introduced by Aumann (1964)].

2.12. Assignments

(XIII) *An assignment* $\boldsymbol{a} : I \to \Omega$ *is a function such that for every* $V \in \mathscr{B}(K)$, $t \to \boldsymbol{a}(t)(V)$ *is measurable*[2] *or, equivalently (Fact 1 in section 3.1), a weak-star measurable function [i.e.,* $t \mapsto f\boldsymbol{a}(t)$ *is measurable for every* $f \in C(K)$*].*

Given an assignment $\boldsymbol{a} : I \to \Omega$ with

$$\int_I \boldsymbol{a}(t)(h)\, d\lambda(t) < \infty,$$

the *aggregate commodity distribution* corresponding to $\boldsymbol{a}$, denoted $\int_I \boldsymbol{a}\, d\lambda$, is the element of $m(K)$ defined by

$$\left(\int_I \boldsymbol{a}\, d\lambda\right)(V) = \int_I \boldsymbol{a}(t)(V)\, d\lambda(t) \quad \text{for every} \quad V \in \mathscr{B}(K)$$

(see Fact 2 in section 3.1). If we write $\int_I \boldsymbol{a}\, d\lambda$, we are implicitly asserting

$$\int_I \boldsymbol{a}(t)(h)\, d\lambda(t) < \infty.$$

(XIV) *An allocation for an economy* ν *is an assignment* $\boldsymbol{a}$ *such that* $\int_I \boldsymbol{a}\, d\lambda \le \int_\Omega i\, d\nu_E$.

2. Hence an assignment is a random measure, with range measures of a very special kind.

2.13. Representation of Distributions

(XV) *A probability measure* τ *on* $P \times E \times \Omega$ *is represented by* $e : I \to P \times E$ *and* $\boldsymbol{a} : I \to \Omega$ *if* $\tau = \lambda \circ (e, \boldsymbol{a})^{-1}$.

Since $P \times E \times \Omega$ is a complete metric space, any distribution on $P \times E \times \Omega$ has a representation.

2.14. Core Distributions

(XVI) *A probability measure* τ *on* $P \times E \times \Omega$ *belongs to the core of an economy* ν *if* $\tau_{P\times E} = \nu$, $\int i \, d\tau_\Omega \leq \int i \, d\tau_E$, *and, given a representation,* $e : I \to P \times E$, $\boldsymbol{a} : I \to \Omega$ *for* τ, *there is no measurable set* $A \subset I$ *and an assignment* $\boldsymbol{a}' : I \to \Omega$ *such that*

(i) $\lambda(A) > 0$;
(ii) $\int_A \boldsymbol{a}' \, d\lambda \leq \int_A \omega \, d\lambda$;
(*iii*) $\boldsymbol{a}'(t) \succ_\tau \boldsymbol{a}(t)$ *for a.e.* $t \in A$

[*where* $e(t) = (\succsim_\tau, \omega(t))$].

A byproduct of the theorem about to be stated (in section 2.15) is the *independence* of the previous definition of the particular representation for τ used.

2.15. Equivalence Theorem

Theorem 2. *A distribution* τ *on* $P \times E \times \Omega$ *belongs to the core of an economy* ν *if and only if it is an equilibrium distribution for this economy.*

2.16. A Technical Comment

The reader familiar with the continuum of agents' models will have noted that we emphasize distributions over representations, and this in spite of the fact that for the equivalence theorem it is indispensable to consider the economy in representation form.

In the present context, to deal with distributions rather than with representations is not simply a matter of taste, but it makes the existence problem a manageable one. While given an economy ν we can prove the existence of an equilibrium distribution τ (and therefore of an equilibrium allocation $\boldsymbol{a} : I \to \Omega$ for *some* representation of ν, $e : I \to P \times E$), it is very unlikely that given a representation $e : I \to P \times E$, there is an allocation $\boldsymbol{a} : I \to \Omega$ which is an equilibrium with respect to this e. Of course, the problem is not one of economic substance, it merely revolves on measurability technicalities. The main difference with the usual set-up, where the commodity space is $\mathbf{R}^l$, is the following: if $F : I \to \Omega$ is a correspondence with measurable graph (and, say, uniformly bounded values) then $\int F \, d\lambda \subset m(K)$ is a well-defined, non-empty set but it may neither be convex nor weak-star closed [it is easy to see, using the Liapunov theorem on the range of vector measures into $\mathbf{R}^n$, that its weak-star *is* convex; see Kluvánek (1974)].

3. The Proofs

3.1. *Mathematics*

In this section we gather together a number of mathematical results which shall be repeatedly used later on and are not explicitly found in, say, Hildenbrand (1974).

For the rest of this (and only this) section the following notational conventions are maintained: M, L are complete, metric spaces; $\mathscr{B}(M)$ is the Borel σ-field of M; $C(M)$, $m(M)$, $m^+(M)$ are, respectively, the set of continuous functions, Borel measures, and non-negative Borel measures on M; μ is a generic symbol for a measure; $(Z, \mathscr{Z}, \rho)$ is an abstract measure space. For measures on a compact M to converge weakly we mean the usual weak-star convergence.

Fact 1. Let $g : Z \to m(M)$, then $t \mapsto g(t)(V)$ is measurable for every $V \in \mathscr{B}(M)$ if and only if $t \mapsto \int_M f \, dg(t)$ is measurable for every $f \in C(M)$.

Proof. Use the exercise on σ-additive classes of sets on p. 19 of Neveu (1965).[3] ▪

Fact 2. Let $g : Z \to m^+(M)$; if $t \mapsto g(t)(V)$ is measurable for every $V \in \mathscr{B}(M)$ and $\int_Z g(t)(M) \, d\rho(t) < \infty$, then $\int_Z g(t)(V) \, d\rho(t)$ defines a measure [i.e., an element of $m^+(M)$], denoted $\int_Z g \, d\rho$, which coincides with

3. I am extremely indebted to F. Delbaen, not only for indicating to me this proof, but for persuading me that Fact 1 was true when I had already given up.

the measure induced (via the Riesz representation theorem) by the linear functional on $C(M)$,

$$f \mapsto \int_Z \left(\int_M f \, \mathrm{d}g(t)\right) \mathrm{d}\rho(t).$$

Proof. Easy. ∎

Fact 3. Let M, L be compact and $g : M \to m^+(L)$ be weak-star continuous. If $\mu_n, \mu \in m(M)$ and $\mu_n \to \mu$ weakly, then $\int_M g \, \mathrm{d}\mu_n \to \int_M g \, \mathrm{d}\mu$ weakly.

Proof. Let $f \in C(L)$. It is easily checked that

$$\int_L f \, \mathrm{d}\left(\int_M g \, \mathrm{d}\mu_n\right) = \int_M \left(\int_L f \, \mathrm{d}g(t)\right) \mathrm{d}\mu_n(t).$$

Since g is weak-star continuous, $\int_L f \, \mathrm{d}g(t)$ is a continuous function of t; hence

$$\int_M \left(\int_L f \, \mathrm{d}g(t)\right) \mathrm{d}\mu_n(t) \to \int_M \left(\int_L f \, \mathrm{d}g(t)\right) \mathrm{d}\mu(t) = \int_L f \, \mathrm{d}\left(\int_M g \, \mathrm{d}\mu\right).$$

So, $\int_M g \, \mathrm{d}\mu_n \to \int_M g \, \mathrm{d}\mu$ weakly. ∎

Fact 4. Let M be compact, $t \in M$, and $\mu \in m^+(M)$. Then, for any $\varepsilon > 0$, there are pairwise disjoint open sets $B_i \subset M$, $1 \leq i \leq n$, such that

$$\mu\left(\bigcup_{i=1}^{n} B_i\right) = \mu(M), \qquad t \in \bigcup_{i=1}^{n} B_i,$$

and radius $B_i < \varepsilon$ for every i.

Proof. Easy. ∎

Fact 5. Let $V \subset [0, 1]$ be measurable. If $\mu \in m(M)$ and $\mu(M) = \lambda(V)$, then $\mu = \lambda \circ \varphi^{-1}$ for some measurable $\varphi : V \to M$ (where λ is Lebesgue measure).

Proof. There is a function φ_1: $[0, \mu(M)] \to$ M with $\mu = \lambda \circ \varphi_1^{-1}$ [see Hildenbrand (1974, p. 50)]; there is also a function φ_2: $V \to [0, \mu(\mathrm{M})]$ with $\lambda = \lambda \circ \varphi_1^{-1}$ [see Gabszewicz and Mertens (1971, p. 715)]. Take $\varphi = \varphi_1 \circ \varphi_2$. ∎

Fact 6. Let M be compact and $M_n \subset M$ closed; if $M_n \to M$ in closed convergence and $\mu \in m(M)$, then we can find $\mu_n \in m(M_n)$ such that $\mu_n \to \mu$ weakly. If μ is integer-valued, non-negative, and $\mu(M) \leq \delta$, then μ_n can be chosen to have those properties.

Proof. Easy. ∎

Fact 7. Let M be compact and $M_n \subset M$ closed. If $M_n \to M$ in closed convergence and $\mu_n \in m(M)$ is such that $\mu_n \to \mu$ weakly and $\mu_n(M_n) = 1$, then $\mu(M) = 1$.

Proof. Immediate. ■

3.2. Preliminaries

N.B.: *For the rest of section 3, the symbols K, Ω, $\mathscr{P}$, P, ξ, E, H_p will stand for the particular entities introduced in section 2; ν is the fixed economy under consideration. Except for ν, τ, ω, and λ small-type Greek letters always denote real numbers;* $\mathbf{R} = (-\infty, \infty)$. *Only measurable subsets of I are considered.*

Without loss of generality (w.l.o.g.) we can assume that E has the form

$$E = \{a \in \Omega : a(h) \leq \beta\}.$$

Define

$$C_1^+(K) = \{p \in C^+(K) : p(h) = 1\};$$

$C_1^+(K)$ will be a space of normalized prices, the h commodity being a numeraire.

Let $\mathscr{C}(K)$ be the space of compact subsets of K which include h [i.e., $\mathscr{C}(K) = \{J \subset K$: J is closed, $\{h\} \subset J\}$]; $\mathscr{C}(K)$ endowed with the closed convergence topology becomes a compact, metrizable space. For $J \in \mathscr{C}(K)$, denote $J^d = J \sim \{h\}$.

Remember Ω has the form

$$\Omega = \{a \in m(K) : a \geq 0,\ a \mid K^d \text{ is integer-valued, } a(K^d) \leq \alpha\};$$

for every $J \in \mathscr{C}(K)$ define

$$\Omega_J = \{a \in m(K) : a \geq 0,\ a \mid J^d \text{ is integer-valued, } a(J^d) \leq \alpha\}.$$

By identifying $a \in \Omega_J$ to the element $a' \in \Omega$ with $a' \mid J = a$ and $a'(K/J) = 0$, Ω_J can be regarded as a subset of Ω [i.e., $\Omega_J = \{a \in \Omega : a(K/J) = 0\}$]. Define then

$$\mathscr{P}_J = \{\succsim \cap\, \Omega_J \times \Omega_J : \succsim\, \in \mathscr{P}\};$$

clearly, $\mathscr{P}_J$ can be identified with the set determined by the properties defining $\mathscr{P}$ except that K is replaced throughout by J. Let

$$P_J = \{\succsim \cap \Omega_J \times \Omega_J : \succsim \in P\},$$

$$E_J = E \cap \Omega_J = \{a \in \Omega_J : a(h) \leq \alpha\},$$

$$\hat{\mathscr{P}} = \bigcup_{J \in \mathscr{P}(K)} \mathscr{P}_J, \qquad \hat{P} = \bigcup_{J \in \mathscr{C}(K)} P_J;$$

$\hat{\mathscr{P}}$ and $\hat{P}$ are endowed with the closed convergence topology. We define $C_1^+(J)$, for $J \in \mathscr{C}(K)$, analogously to $C_1^+(K)$ and will freely identify $f \in C_1^+(J)$ with the function f' defined on K *by* $f' \mid J = f$ and $f'(s) = 0$ for $s \notin J$.

Lemma 1. The set $\{(a, a', \succsim) \in \Omega \times \Omega \times \hat{\mathscr{P}} : a \succsim a'\}$ *is closed (in the product topology).*

Proof. If $a_n \to a$, $a'_n \to a'$, $\succsim_n \to \succsim$, $a_n \succsim_n a'_n$, then the sequence of sets $\succsim_n$ does eventually intersect every neighborhood of $(a, a') \in \Omega \times \Omega$, hence $(a, a') \in \text{Ls}\,(\succsim_n) = \succsim$.[4] ■

Lemma 2. The function $\Psi : \mathscr{C}(K) \times \mathscr{P} \to \hat{\mathscr{P}}$ *given by* $(J, \succsim) \to \succsim \cap \Omega_J \times \Omega_J$ *is continuous.*

Proof. Let $\succsim_n \to \succsim$, $J_n \to J$.

$$\text{Ls}(\succsim_n \cap \Omega_{J_n} \times \Omega_{J_n}) \subset \succsim \cap \Omega_J \times \Omega_J$$

follows from Lemma 1. Let $a' \succsim a$, $a', a \in \Omega_J$. For arbitrary $\varepsilon > 0$, put $a'' = a' + \varepsilon b$; then $a'' \succ a$ and, by Fact 6, there are $a''_n \to a''$, $a_n \to a$, with $a_n, a''_n \in \Omega_{J_n}$. By Lemma 1 there is N such that if $n > N$, then $a''_n \succsim_n a_n$, i.e.,

$$(a''_n, a_n) \in \succsim_n \cap \Omega_{J_n} \times \Omega_{J_n}.$$

Since a'' is arbitrarily close to a', this proves

$$\text{Li}\,(\succsim_n \cap \Omega_{J_n} \times \Omega_{J_n}) \subset \succsim \cap \Omega_J \times \Omega_J. \quad ■$$

Lemma 3. $\hat{P}$ *is compact.*

Proof. Since $\mathscr{C}(K)$ and P are compact spaces, it suffices to prove that $J_n \to J$, $\succsim_n \to \succsim$, $\succsim_n \in P$, implies

$$\succsim_n \cap \Omega_{J_n} \times \Omega_{J_n} \to \succsim \cap \Omega_J \times \Omega_J,$$

but this is the content of Lemma 2. ■

4. For the definition of Li and Ls, see Hildenbrand (1974, p. 15).

We now define a notion of convergence for pairs $J \in \mathscr{C}(K)$, $p \in C_1^+(J)$. Let $J_n, J \in \mathscr{C}(K)$, $p_n \in C_1^+(J_n)$, $p \in C_1^+(K)$; then we write $(J_n, p_n) \to (J, p)$ if and only if $J_n \to J$ and, for every subsequence (J_{n_i}, p_{n_i}) one has: $s_{n_i} \to s$, $s_{n_i} \in J_{n_i}$, implies $p_{n_i}(s_{n_i}) \to p(s)$. From now on, if we write (J, p), we imply $J \in \mathscr{C}(K)$ and $p \in C_1^+(J)$

Lemma 4. If $(J_n, p_n) \to (K, p)$ *and* $a_n \in \Omega_{J_n}$, $a_n \to a$, *then* $p_n a_n \to pa$.

Proof. Since $a_n \to a$, we have $pa_n \to pa$. Let $\varepsilon > 0$ be arbitrary. Because $(J_n, p_n) \to (K, p)$ there is N such that if $n > N$, then

$$\sup_{s \in J_n} |p_n(s) - p(s)| \leq \varepsilon;$$

we can also assume that $n > N$ implies $a_n(K) \leq a(K) + \varepsilon$; hence, for $n > N$,

$$|p_n a_n - pa_n| < \varepsilon(a(K) + \varepsilon).$$

So, $p_n a_n \to pa$. ■

Analogously to section 2, define, for every $J \in \mathscr{C}(K)$ and $p \in C_1^+(J)$,

$$\begin{aligned} H(J, p) = \{(\precsim, \omega, a) \in \hat{P} \times \Omega \times \Omega:\ \ \precsim \in P_J;\ \ \omega, a \in \Omega_J;\\ pa \leq p\omega;\ \ \text{and}\ \ a' \in \Omega_J,\ \ pa' \leq p\omega\ \ \text{implies}\ \ a \succsim a'\}.\end{aligned}$$

Lemma 5. If $(J_n, p_n) \to (K, p)$, *then* $\text{Ls}(H(J_n, p_n)) = H(K, p)$. *Moreover, for every* (J, p), $H(J, p) \neq \phi$.

Proof. The second part follows from the (weak-star) compactness of $\{a \in \Omega_J : pa \leq \gamma\}$.

To prove the first part, it suffices to show that

$$(\precsim_n, \omega_n, a_n) \in H(J_n, p_n),$$

$$(\precsim_n, \omega_n, a_n) \to (\precsim, \omega, a) \in \mathscr{C}(\Omega \times \Omega) \times \Omega \times \Omega$$

implies

$$(\precsim, \omega, a) \in H(K, p).$$

By Lemma 3, $\precsim \in \hat{P}$ and, combining $J_n \to K$ with Fact 6 and Lemma 1 we easily deduce that $\precsim$ is complete, i.e., $\precsim \in P = P_K$.

Let $pa' \leq p\omega$. If $a'(h) = 0$, then $a \succsim a'$ by the definition of preferences. If $a'(h) > 0$, then $pa' > 0$ and, in order to prove $a \succsim a'$, we can as well assume $pa' \leq p\omega$. By Fact 6, there is $a'_n \in J_n$, $a'_n \to a$. By Lemma 4,

$p_n a'_n \to pa'$ and $p_n \omega_n \to p\omega$. Hence there is N such that if $n > N$, then $p_n a'_n < p_n \omega_n$; this implies $a_n \succsim_n a'_n$ and, by Lemma 1, $a \succsim a'$. Therefore, $(\succsim, \omega, a) \in H(K, p)$. ■

Lemma 6. *Let* $\succsim_n \in P_{J_n}$, $J_n \in \mathscr{C}(K)$. *If* $w_n \to \infty (w_n \in R)$, $p_n \in C_1^+(J_n)$ *and* $a_n \in \Omega$ *is such that "*$a \in \Omega_{J_\eta}$ *and* $p_n a \le w_n$ *implies* $a_n \succsim_n a$*", then* $a_n(h) \to \infty$.

Proof. Suppose not. Then we can assume $a_n \to a \in \Omega$. By Lemma 3, $\succsim_n \to \succsim \in \hat{P}$. By property (iv) in the definition of preferences there is $\gamma > 0$ such that $\gamma b \succ a$. But, for n large enough (use Lemma 1),

$$\gamma p_n b = \gamma < w_n \quad \text{and} \quad \gamma b \succ_n a_n.$$

Contradiction. ■

3.3. Existence of Equilibria in the Finite Number of Commodities Case

We now prove the existence of equilibria for economies having a finite number of commodities. Except for the indivisibilities aspect, everything is as in Aumann (1966), Schmeidler (1969) or Hildenbrand (1974). Indivisibilities could potentially disturb the upper hemicontinuity (u.h.c.) of the individual excess demand correspondences. However, with our strong assumptions, we have already proved that they are indeed u.h.c. (Lemma 5). So, in this section, we shall be rather brief.

Lemma 7. *Suppose that* $J \in \mathscr{C}(K)$ *is a finite set and* ν' *is an economy on* $P_J \times E_J$*, then there is an equilibrium distribution* τ' *(on* $P_J \times E_J \times \Omega_J$*) for* ν'.

Proof. Let $l = \#(J)$ and Z be the non-negative integers. For economy of notation we drop all the subindexes J. We identify Ω with

$$\left\{ a \in Z^{l-1} : \sum_{i=1}^{l-1} a^i \le \alpha \right\} \times [0, \infty)$$

and let

$$\Delta = \left\{ q \in \mathbf{R}^l : q^i \ge 0 \text{ for all } i,\ q^l > 0,\ \sum_{i=1}^{l} q^i = 1 \right\},$$

(i.e., we are changing the price normalization). Define the correspondence

$$g : P \times \Delta \times (0, \infty) \to \Omega$$

by

$$g(\succsim, p, w) = \{a \in \Omega : pa \leq w, \quad \text{and} \quad pa' \leq w \quad \text{implies} \quad a \succsim a'\},$$

i.e., g is the demand correspondence. By Lemma 5, g is u.h.c. By the monotonicity with respect to h, $pg(\succsim, p, w) = w$. By Lemma 6, if $\gamma > 0$, $\succsim_n \in P$, $p_n \in \Delta$, $w_n \geq \gamma$, $a_n \in g(\succsim_n, p_n, w_n)$ and $p_n^l \to 0$, then $a^l \to \infty$.

Let $e : I \to P \times E$ be a representation for ν'. Then, writing $e(t) = (\succsim_t, \omega(t))$,

$$\textstyle\int_I \omega \, d\lambda \gg 0.$$

Define the mean excess demand correspondence $G : \Delta \to \mathbf{R}^l$ as the (set-valued) integral

$$G(p) = \textstyle\int_I (g(\succsim_t, p, p\omega_t) - \{\omega(t)\}) \, d\lambda(t).$$

Using the previous observations, it is seen (as in the references above) that G can indeed be defined and has the following properties:

(i) there is $\gamma > 0$ such that $G(p) > -(\gamma, \ldots, \gamma)$ for all p;
(ii) $pG(p) = 0$;
(iii) G is u.h.c. and convex-valued;
(iv) if $p_n \in \Delta$, $p_n^l \to 0$ and $x_n \in G(p_n)$, then $x_n^l \to \infty$.

The existence of an equilibrium distribution for ν' is equivalent to the existence of $p \in \Delta$ and $x \in G(p)$ such that $x \leq 0$. Such a pair (p, x) exists for a function G satisfying the properties transcribed in the previous paragraph. [Proof: let

$$\Delta_\delta = \{q \in \Delta : q^l \geq \delta\}$$

with δ small enough to guarantee

$$`q^l = \delta, \quad x \in G(q) \quad \text{implies} \quad x^l > l\gamma'.$$

As in Mas-Colell (1974) there is $p \in \Delta_\delta$ and $x \in G(p)$ such that $qx \leq 0$ for all $q \in \Delta_\delta$; this is immediately seen to imply $x \leq 0$.] ■

3.4. *Equicontinuity*

Let $J_n \in \mathscr{C}(K)$ and $p_n \in C_1^+(J_n)$. The sequence (J_n, p_n) will be called *equicontinuous* if for every $\varepsilon > 0$ there is $\delta > 0$ such that, for all n, if $s, s' \in J_n$ and $d_K(s, s') \leq \delta$, then

$$|p_n(s) - p_n(s')| \leq \varepsilon.$$

Lemma 8. *Let $J_n \in \mathscr{C}(K)$, $p_n \in C_1^+(J_n)$, $J_n \to K$ and $J_n \subset J_{n+1}$ (all n). If the sequence (J_n, p_n) is equicontinuous and, for some η, $p_n(s) \leq \eta$ for all n and $s \in K$, then there is $p \in C_1^+(K)$ such that $(J_{n_i}, p_{n_i}) \to (J, p)$ for a subsequence (J_{n_i}, p_{n_i}) of (J_n, p_n).*

Proof. This follows by familiar arguments in analysis. First, replace every J_n by a finite $J'_n \subset J_n$ preserving

$$J'_n \subset J'_{n+1} \quad \text{and} \quad J'_n \to K;$$

$$J' = \bigcup_n J'_n$$

is countable and dense in K. Applying a diagonal argument, we can assume (extracting a subsequence if needed) that there is $p': J' \to \mathbf{R}$ with $p_n(s) \to p'(s)$ for all $s \in J'$ [see Royden (1969, 9.30, p. 177)]. Of course, $p'(h) = 1$.

We show that p' defines a continuous function on K, i.e., $d_K(s_m, s'_m) \to 0$ and $s_m, s'_m \in J'$ implies

$$p'(s_m) - p'(s'_m) \to 0\,.$$

Suppose otherwise; extracting subsequences if needed, we can assume

$$d_K(s_m, s'_m) \to 0$$

and

$$\lim\,(p'(s_m) - p'(s'_m)) > \varepsilon > 0.$$

Let $\delta > 0$ correspond to ε as in the definition of equicontinuity and pick M such that

$$d_K(s_M, s'_M) < \delta \quad \text{and} \quad p'(s_M) - p'(s'_M) > \varepsilon.$$

We have a contradiction since then for N large enough $s_M, s'_M \in J'_N$ and $p_N(s_M) > p_N(s'_M) + \varepsilon$. Hence, p' defines a continuous function on K which we shall call p. A repetition of the argument above yields $(J'_n, p_n) \to (K, p)$ and $(J_n, p_n) \to (K, p)$—remember also that at this point (J_n, p_n) needs only be a subsequence of the original sequence. ■

We now become more specific and construct a particular sequence $K_n \in \mathscr{C}(K)$, $K_n \to K$.

By Fact 4, we can find $\delta_n \to 0$ and, for every n, a collection of points $s_i^n \in K$, $1 \leq i \leq m_n$, and pairwise disjoint neighborhoods of s_i^n, $B_i^n \subset K$, $1 \leq i \leq m_n$, such that, denoting

$$B^n = \bigcup_{i=1}^{m_n} B_i^n, \qquad K_n = \bigcup_{i=1}^{m_n} \{s_i^n\} :$$

(i) $d_K(s_i^n, s) \le \delta_n$ for every $s \in B_i^n$ and all n, $1 \le i \le m_n$;
(ii) $(\int_\Omega i \, \mathrm{d}\nu_E)(B^n) = (\int_\Omega i \, \mathrm{d}\nu_E)(K)$ for all n;
(iii) for every n and $1 \le i \le m_{n}$, if $n' > n$, then $B_i^n \subset B_j^{n'}$ for some $1 \le j \le m_{n'}$;
(iv) $B_1^n = \{h\}$ and $K_n \subset K_{n+1}$ for all n.

Observe that $K_n \to K$ and $K_n \in \mathscr{C}(K)$.

Let $e : I \to P \times E$ be a from now on fixed, but otherwise indeterminated, representation for ν. Denote $\Omega_{K_n}, P_{K_n}, E_{K_n}$ by Ω_n, P_n, E_n and put

$$e_n(t) = \left(\succsim_t^n, \boldsymbol{\omega}_n(t)\right).$$

Define

$$e_n : I \to P_n \times E_n$$

as follows: $\succsim_t^n$ is the restriction of $\succsim_t$ to K_n (i.e., $\succsim_t^n = \succsim_t \cap K_n \times K_n$), $\boldsymbol{\omega}_n(t)$ is the measure determined by

$$\boldsymbol{\omega}_n(t)(K/K_n) = 0$$

and

$$\boldsymbol{\omega}_n(t)\left(s_i^n\right) = \boldsymbol{\omega}(t)\left(B_i^n\right), \qquad 1 \le i \le m_n;$$

clearly $\boldsymbol{\omega}_n(t)$ belongs to E_n and e_n is measurable (use Lemma 2). Of course,

$$\left(\int_I \boldsymbol{\omega}_n \, \mathrm{d}\lambda\right)\left(s_i^n\right) = \left(\int_I \boldsymbol{\omega} \, \mathrm{d}\lambda\right)\left(B_i^n\right)$$

and, since

$$(\int_I \boldsymbol{\omega} \, \mathrm{d}\lambda)(B) = (\int_\Omega i \, \mathrm{d}\nu_E)(B) = (\int_\Omega i \, \mathrm{d}\nu_E)(K) = (\int_I \boldsymbol{\omega} \, \mathrm{d}\lambda)(K),$$

we have

$$(\int_I \boldsymbol{\omega}_n \, \mathrm{d}\lambda)(K_n) = (\int_I \boldsymbol{\omega} \, \mathrm{d}\lambda)(K).$$

Consider any nonempty, open set $V \subset K^d$. Since

$$\operatorname{supp}\left(\int_I \boldsymbol{\omega} \, \mathrm{d}\lambda\right) = K,$$

there is, by the construction of e_n, a $\gamma_V > 0$ such that

$$\lambda\{t : \boldsymbol{\omega}_n(t)(V) > 0\} > \gamma_V \quad \text{for all} \quad n.$$

Lemma 9. *Let* $\varepsilon_n \to 0$ *and suppose that* $K'_n, K''_n \in \mathscr{C}(K)$, $A'_n, A''_n \subset I$, $p_n \in C_1^+(K_n)$, *and* $\boldsymbol{a}_n\colon I \to \Omega_n$ *satisfy, for all n, the following properties:*

(i) $K''_n \subset K'_n \subset K_n$ *and* $K''_n \to K$;
(ii) $A'_{n+1} \subset A'_n$, $A''_{n+1} \subset A''_n$, $A'_n \subset A''_n$, *and* $\lambda\left(A''_n\right) \to 0$;
(iii) $\int_I \boldsymbol{a}_n \, d\lambda = \int_I \omega_n \, d\lambda$;
(iv) *for a.e.* $t \in I/A'_n$, $p_n\boldsymbol{a}_n(t) \le p_n\boldsymbol{\omega}_n(t) + (1/n)$ *and* "$p_n a \le p_n\boldsymbol{\omega}_n(t) - 2\varepsilon_n$, $a \in \Omega_n$, *implies* $\boldsymbol{a}_n(t) + \varepsilon_n b \succsim_t a$";
(v) *if* $s \in K'_n$, *then* $\left(\int_{I/A_{n'}} \boldsymbol{a}_n\right)(\{s\}) > 0$; *if* $s \in K''_n$, *then* $\left(\int_{I/A_{n''}} \boldsymbol{a}_n\right)(\{s\}) > 0$;
(vi) *if* $s \in K_n/K'_n$, *then* $\left(\int_{I/A_{n''}} \boldsymbol{\omega}_n\right)(\{s\}) = 0$.

Then $\left\{\sup_{s\in K_{n'}} p_n(s)\right\}_{n=1}^{\infty} < \eta$ *for some* η *and* $\left(K''_n, p_n \mid K''_n\right)$ *is equicontinuous.*

Proof. For any $s, s' \in K^d$ and $a \in \Omega$ with $a(s) > 0$, we define $a' \in \Omega$ by letting a' be equal to a except that one unit of s is replaced by one unit of s'; for the rest of the proof the symbols a, a' will always stand in this relation, the corresponding s and s' will be clear by the context.

According to condition (iii) in the definition of consumer characteristics,

$$a(h) = \hat{a}(h) \quad \text{and} \quad d'_\Omega(a \mid K^d, \hat{a} \mid K^d) < 1/\xi$$

implies

$$a + \xi b \succ \hat{a} \quad \text{for all} \quad a, \hat{a} \in \Omega \quad \text{and} \quad \succsim \in P.$$

Since $\{a \mid K^d : a \in \Omega\} \subset m(K^d)$ is compact, we can assume (by putting the ξ constant large enough) that, for some N,

$$d_K(s, s') < 1/\xi$$

implies

$$a + \xi b \succ a' + \varepsilon_n b \quad \text{for all} \quad n > N,\ s, s' \in K, \quad \text{and} \quad \succsim \in P.$$

We show first that

$$\left\{\sup_{s\in K_{n'}} p_n(s)\right\}_{n=1}^{\infty} < \eta \quad \text{for some } \eta.$$

Clearly, for every non-empty, open $V \subset K^d$ and n large enough, we have

$$\lambda\left\{t \in I/A'_n : p_n \boldsymbol{\omega}_n(t) \geq \inf_{s \in V} p_n(s)\right\} > \frac{\gamma v}{2}.$$

Suppose there is a non-empty, open set $V \subset K^d$ for which

$$\limsup \inf_{s \in V} p_n(s) = \infty;$$

then, using Lemma 6 and Fatou's lemma, we have

$$\limsup \left(\int_{I/A_n} \boldsymbol{a}_n \, d\lambda \right)(h) = \infty$$

which contradicts assumption (iii). Hence, for every non-empty, open $V \subset K^d$,

$$\left\{ \inf_{s \in V} p_n(s) \right\}_{n=1}^{\infty}$$

is bounded.

Suppose that

$$\left\{ \sup_{s \in K_{n'}} p_n(s) \right\}_{n=1}^{\infty}$$

is not bounded. We can assume we have $s_n \in K'_n$, $s_n \to \bar{s}$, $p_n(s_n) \to \infty$. Let $B^n_{1/2\xi}(\bar{s})$ be the $1/2\xi$ neighborhood of $\bar{s}$ in K'_n; then,

$$\inf_{s \in B_{1/2\xi}(\bar{s})} p_n(\bar{s}) < \theta \quad \text{for all } n \text{ and some } \theta.$$

Pick $n > N$ large enough to have

$$p_n(s_n) - p_n(s'_n) > \xi + 2\varepsilon_n = 1/n \quad \text{for some} \quad s'_n \in B^n_{1/2\xi}(\bar{s}).$$

This yields a contradiction since, by assumption (v),

$$\lambda(\{t \in I/A'_n : \boldsymbol{a}_n(t)(s_n) > 0\}) > 0,$$

and if $\boldsymbol{a}_n(t)(s_n) > 0$, then

$$p_n(\boldsymbol{a}'_n(t) + \xi b) \leq p_n \boldsymbol{a}_n(t) - 2\varepsilon_n - \frac{1}{n} \leq p_n \boldsymbol{\omega}_n(t) - 2\varepsilon_n$$

and

$$\boldsymbol{a}'_n(t) + \xi b \succ_t \boldsymbol{a}_n(t) + \varepsilon_n b.$$

Now we show equicontinuity of $(K_n'', p_n \mid K_n'')$ by a similar argument. Let $(K_n'', p_n \mid K_n'')$ not be equicontinuous. Then we can assume we have $d_K(s_n, s_n') \to 0$ and

$$\lim p_n(s_n) > \lim p_n(s_n') + \theta \quad \text{for} \quad s_n, s_n' \in K_n'' \quad \text{and some } \theta.$$

Combining assumptions (iv) and (vi) it follows that for a.e. $t \in I/A_n''$ and n sufficiently large,

$$\boldsymbol{a}_n(t) \in \Omega' = \{a \in \Omega : a(h) \leq \beta\eta + 1\}.$$

By the compactness of $\hat{P}$ and Ω' and the desirability of h there is $\mu > 0$ and N' such that if $\succsim \in \hat{P}$, a, $\hat{a} \in \Omega'$, $d_\Omega(a, \hat{a}) \leq \mu$ and $n > N'$, then

$$a + \frac{\theta}{2}b \succ \hat{a} + \varepsilon_n b.$$

But this yields again a contradiction since for s_n sufficiently close to s_n' and $n > N'$ sufficiently large we will have

$$\lambda(\{t \in I/A_n'' : \boldsymbol{a}_n(t)(s_n) >\}) > 0,$$

and, if $\boldsymbol{a}_n(t)(s_n) > 0$, then

$$\boldsymbol{a}_n'(t) + \frac{\theta}{2}b \succ_t \boldsymbol{a}_n(t) + \varepsilon_n b$$

and

$$p_n\left(\boldsymbol{a}_n'(t) + \frac{\theta}{2}b\right) \leq p_n\boldsymbol{a}_n + \frac{\theta}{2} - \theta \leq p_n\omega_n(t) + \frac{1}{n} - \frac{\theta}{2} \leq p_n\omega_n(t) - 2\varepsilon_n. \quad \blacksquare$$

3.5. *Proof of the Existence Theorem (Theorem 1)*

Let $\nu^n = \lambda \circ e_n^{-1}$; ν^n; ν^n is an economy on $P_n \times E_n$; by Lemma 7 there are equilibrium prices p_n and distributions τ^n for ν^n. W.l.o.g. (remember that preferences are monotone) we assume

$$\textstyle\int_\Omega i \, d\tau_\Omega^n = \int_\Omega i \, d\nu_E^n.$$

With $\varepsilon_n = 0$, $K_n' = K_n'' = K_n$, $A_n' = A_n'' = \phi$ and $\boldsymbol{a}_n$ such that

$$\tau^n = \lambda \circ (e_n, \boldsymbol{a}_n)^{-1}$$

[such $\boldsymbol{a}_n$ does exist; remember $\#(K_n) < \infty$], the sequence (K_n, p_n) satisfies the conditions of Lemma 9, hence by Lemmas 8 and 9 we can assume $(K_n, p_n) \to (K, p)$ for some $p \in C_1^+(K)$.

Pick

$$\eta > \max_{s \in K}, p(s)$$

and define

$$\Omega' = \{a \in \Omega : a(h) \leq \eta\beta + 1\};$$

then Ω' is compact and, using Lemma 5, so are $H(K_n, p_n)$ and $H(K, p)$.

Regarding ν^n as a measure on $\hat{P} \times E$ we have $\nu^n \to \nu$ weakly since (use Lemma 2) $e^n(t) \to e(t)$ for a.e. $t \in I$ [see Hildenbrand (1974, D.39, p. 51)]. For all n sufficiently large, τ_n is a probability on $\hat{P} \times E \times \Omega'$. Because $\hat{P} \times E \times \Omega'$ is compact, we can assume $\tau_n \to \tau$ weakly [see Hildenbrand (1974, D.30, p. 49)]. We claim τ is an equilibrium distribution for ν with prices p. Indeed $\tau^n_{\hat{P}\times E} = \nu^n$, $\nu^n \to \nu$, and $\tau^n_{\hat{P}\times E} \to \tau_{\hat{P}\times E}$ [see W. Hildenbrand (1974, D.27, p. 48)] yields $\tau_{\hat{P}\times E} = \nu$; by Fact 3, $\tau^n_{\Omega'} \to \tau_{\Omega'}$, and $\nu^n_E \to \nu_E$ imply, respectively,

$$\int_\Omega i \, d\tau^n_\Omega \to \int_\Omega i \, d\tau_\Omega \quad \text{and} \quad \int_\Omega i \, d\nu^n_E \to \int_\Omega i \, d\nu_E,$$

i.e.,

$$\int_\Omega i \, d\tau_\Omega = \int_\Omega i \, d\nu_E;$$

finally, by Lemma 5, $\text{Ls}(H(K_n,p_n)) \subset H(K, p)$ and so, $\tau^n(H(K_n, p_n)) = 1$ implies $\tau(H(K, p)) = 1$ (Fact 7).

3.6. *Proof of Theorem 2 (Equivalence)*

The proof that an equilibrium τ is a core distribution is the standard one [see, for example, Debreu and Scarf (1963)]; we skip it.

Let τ be a core distribution and suppose that $\boldsymbol{a} : I \to \Omega$ is an allocation such that $\tau = \lambda \circ (e, \boldsymbol{a})^{-1}$ (remember that e was indeterminate; so, by Fact 5, we can assume that for the given τ such an $\boldsymbol{a}$ exists). W.l.o.g. we assume

$$\int_I \boldsymbol{a} = \int_I \boldsymbol{\omega}.$$

For every n, define $\boldsymbol{a}_n : I \to \Omega_n$ by $\boldsymbol{a}_n(t)(s_i^n) = \boldsymbol{a}(t)(B_i^n)$.

Lemma 10. *Let* $F_n \subset I$, $\boldsymbol{c}_n : F_n \to \Omega_n$,

$$\left(\int_{F_n} \boldsymbol{c}_n\right)(K^d) \leq \left(\int_{F_n} \boldsymbol{\omega}_n\right)(K^d)^5$$

and $\mu > 0$. *Then, for N large enough, if* $n > N$ *there is* $\boldsymbol{c}'_n : F_n \to \Omega$ *such that:*

(i) $d_\Omega(\boldsymbol{c}_n(t), \boldsymbol{c}'_n(t)) < \mu$ *and* $\boldsymbol{c}_n(t)(h) = \boldsymbol{c}'_n(t)(h)$ *for a.e.* $t \in F_n$;

(ii) $\left(\int_{F_n} \boldsymbol{c}'_n\right) \mid K^d \leq \left(\int_{F_n} \boldsymbol{\omega}\right) \mid K^d$.

Proof. Fix an n. For every $1 \leq j \leq \alpha$ and $1 < i \leq m_n$ let

$$F_n^{ij} = \{t \in F_n : \boldsymbol{c}(t)(s_i) = j\}.$$

Then

$$\sum_{j=1}^{\alpha} j\lambda(F_n^{ij}) = \sum_{j=1}^{\alpha} \int_{F_{nij}} \boldsymbol{c}(t)(s_i) \leq (\textstyle\int_{F_n} \boldsymbol{\omega})(B_i^n) = (\textstyle\int_{F_n} \boldsymbol{\omega})(\overline{B}_i^n).$$

Consider a fix i and suppose

$$\left(\int_{F_n} \boldsymbol{\omega}\right)\left(\overline{B}_i^n\right) > 0;$$

let

$$\eta_j = \lambda\left(F_n^{ij}\right) / \left(\int_{F_n} \boldsymbol{\omega}\right)\left(\overline{B}_i^n\right);$$

by Fact 5, there is for every j a measurable map $\varphi_{ij} : F_n^{ij} \to \overline{B}_i^n$ such that

$$\lambda \circ \varphi_{ij}^{-1} = \eta_j (\textstyle\int_{F_n} \omega) \mid \overline{B}_i^n.$$

Define $\boldsymbol{c}'_n : F_n \to \Omega$ as follows: $\boldsymbol{c}'_n(t)(h) = \boldsymbol{c}_n(t)(h)$ and, for $s \in K^d$, $\boldsymbol{c}'_n(t)(\{s\}) = j$ if and only if $s = \varphi_{ij}(t)$ for some i and j. It is easily checked that c'_n is a.e. defined, measurable and

$$\int_{F_n} \boldsymbol{c}'_n \leq \int_{F_n} \boldsymbol{\omega};$$

it is also immediate that $d_\Omega(c_n(t), c'_n(t)) < \mu$ for a.e. $t \in F_n$ and $n > N$ (N large enough). ■

For every n, define $C_n = \{t \in I : \boldsymbol{a}(t)(h) > 2^n\beta\}$. From now on, for $F \subset I$ denote

$$h(F) = \left(\int_F \boldsymbol{\omega}\right)(h).$$

5. For notational conciseness we let $\int_{F_n} c_n = \int_{F_n} c_n \, d\lambda$, etc.

Observe that, for any $F \subset I$, $\lambda(C_n \cap F) \leq h(F)/\beta 2^n$. In particular, $\lambda(C_n) \leq 1/2^n$.

For any $\varepsilon > 0$, $t \in I$, and n let $\varphi_n(\varepsilon, t) = \varepsilon$ if $t \notin C_n$ and $\varphi_n(\varepsilon, t) = \xi$ if $t \in C_n$.

Lemma 11. For any $\varepsilon > 0$ there is N such that if $n > N$ then there is no $F \subset I$ and $\mathbf{c} : F \to \Omega_n$ such that:

(i) $\lambda(F) > 0$;
(ii) $\int_F \mathbf{c} \leq \int_F \boldsymbol{\omega}_n - 2\varepsilon\lambda(F)b$;
(iii) $\mathbf{c}(t) \succ_t \mathbf{a}_n(t) + \varphi_n(\varepsilon, t)b$ *or a.e.* $t \in F$.

Proof. Suppose the lemma is not true. Then we can assume there is $F_n \subset I$ and $\mathbf{c}_n: F_n \to \Omega_n$ such that

$$\lambda(F_n) > 0,$$

$$\int_{F_n} \mathbf{c}_n \leq \int_{F_n} \boldsymbol{\omega}_n - 2\varepsilon\lambda(F_n)b,$$

and

$$\mathbf{c}_n(t) \succ_t \mathbf{a}_n(t) + \varphi_n(\varepsilon, t)b \quad \text{for a.e.} \quad t \in F_n.$$

By Lemmas 2 and 3 and condition (iii) of VIII (section 2), there is $\mu > 0$ sufficiently small such that if, with respect to this μ and c_n, N' and (for $n > N'$) $c'_n : F_n \to \Omega$ satisfy the conclusions of Lemma 10, then:

(i) $\mathbf{c}'_n(t) + \varphi_n(\varepsilon, t)b \succ_t \mathbf{c}_n(t)$ for a.e. $t \in F_n$;
(ii) $(\int_{F_n} \mathbf{c}'_n) \mid K^d \leq (\int_{F_n} \boldsymbol{\omega}) \mid K^d$;
(iii) $(\int_{F_n} \mathbf{c}'_n)(h) + \int_{F_n} \varphi_n(\varepsilon, t)\, d\lambda(t) \leq h(F_n) - 2\varepsilon\lambda(F_n) + \varepsilon\lambda(F_n) + \xi\lambda(F_n \cap C_n)$

$$\leq h(F_n) - \varepsilon\lambda(F_n) + \frac{\varepsilon}{\beta 2^n}\lambda(F_n) \leq h(F_n).$$

Again, by the compactness of $\hat{P}$ and condition (iii) of VIII (section 2), if $N > N'$ is sufficiently large and $n > N$, then

$$\mathbf{a}_n(t) + \varphi_n(\varepsilon, t)b \succ_t \mathbf{a}(t).$$

Summing up, if $n > N$ then

$$\int_{F_n} \mathbf{c}'_n + \left(\int_{F_n} \varphi(\varepsilon, t)\, d\lambda(t)\right)b \leq \int_{F_n} \boldsymbol{\omega}$$

and for a.e. $t \in F_n$,

$$c'_n(t) + \varphi(\varepsilon, t)b \succ_t c_n(t) \succ_t a_n(t) + \varphi_n(\varepsilon, t)b \succ_t a(t).$$

Therefore, $\boldsymbol{a}$ does not induce a core distribution. This contradiction proves the lemma. ■

Lemma 12. *For any $\varepsilon > 0$ there is N such that if $n > N$ then for some $p \in C_1^+(K_n)$, $p\boldsymbol{a} \leq p\boldsymbol{\omega}_n - 2\varepsilon$ implies $\boldsymbol{a}_n(t) + \varphi_n(\varepsilon, t) \succsim_t \boldsymbol{a}$ for a.e. $t \in I$.*

Proof. The content of this lemma is nothing but the usual proof of the equivalence theorem. We follow Hildenbrand (1974, p. 133) and will be rather cursory.

Let N be given by Lemma 11 and take $n > N$. Define the correspondence

$$V : I \to \Omega_n \text{ by } V(t) = (\{a \in \Omega_n : a \succ_t \boldsymbol{a}_n(t) + \varphi_n(\varepsilon, t)b\} \\ -\{\boldsymbol{\omega}_n(t) - 2\varepsilon b\}) \cup \{0\}.$$

Letting $\#(K_n) = l$, we can identify Ω_n with a subset of $\mathbf{R}_+^l$ in the obvious way. For the rest we freely use this identification. Lemma 11 implies

$$0 \notin \textstyle\int_I (V(t) + \mathbf{R}_+^l)\, d\lambda(t)$$

(Note: this is the integral of a correspondence). Since $\int_I (V(t) + \mathbf{R}_+^l)\, d\lambda(t)$ is convex we can let $p \in \mathbf{R}_+^l$ be such that $px \geq 0$ for all

$$x \in \textstyle\int_I (V(t) + \mathbf{R}_+^l)\, d\lambda(t).$$

It is easily checked that $p(h) > 0$, so we can take $p \in C_1^+(K_n)$. Then, for a.e. $t \in I$, $a \succ_t \boldsymbol{a}_n(t) + \varphi_n(\varepsilon, t)b$ implies $pa \geq p\omega_n(t) - 2\varepsilon$ and since we should have $a(h) > 0$ the continuity of $\succsim_t$ and the desirability of h yields that the last inequality holds strictly. ■

Let $\mu > 0$ and i be any positive integer. Pick $\varepsilon > 0$ such that when N is given by Lemma 12 we have

$$\frac{\mu}{2^i} > 3\varepsilon + \frac{\xi}{2^N}.$$

For $n > N$ pick $p_n \in C_1^+(K_n)$ satisfying the conclusions of Lemma 12; then it follows that, for a.e. $t \in I$,

$$p_n\boldsymbol{a}_n(t) + \varphi(\varepsilon, t) \geq p_n\boldsymbol{\omega}_n(t) - 2\varepsilon,$$

and therefore, for any $F \subset I$,

$$\begin{aligned} \textstyle\int_F p_n \boldsymbol{a}_n &\geq \textstyle\int_F p_n \boldsymbol{\omega}_n - 2\varepsilon - \int_F \varphi(\varepsilon, t)\, d\lambda(t) \\ &\geq \textstyle\int_F p_n \boldsymbol{\omega}_n - 2\varepsilon - \xi\lambda(C_n \cap D) - \varepsilon \\ &\geq \textstyle\int_F p_n \boldsymbol{\omega}_n - 3\varepsilon - \frac{\xi}{2^n} > \int_F p_n \boldsymbol{\omega}_n - \frac{\mu}{2^i}. \end{aligned}$$

For every $n > N$ denote

$$D_n = \{t \in I : p_n \boldsymbol{a}_n(t) \geq p_n \boldsymbol{\omega}_n(t) + \mu\}.$$

Suppose that $\lambda(D_n) > 1/2^i$; then

$$\begin{aligned} \textstyle\int_I p_n \boldsymbol{a}_n &= \textstyle\int_{D_n} p_n \boldsymbol{a}_n + \int_{I/D_n} p_n \boldsymbol{a}_n > \int_{D_n} p_n \boldsymbol{\omega}_n + \frac{\mu}{2^i} + \int_{I/D_n} p_n \boldsymbol{\omega}_n - \frac{\mu}{2^i} \\ &= \textstyle\int_I p_n \boldsymbol{\omega}_n = \int_I p_n \boldsymbol{a}_n, \end{aligned}$$

which is a contradiction. Hence, $\lambda(D_n) < 1/2^i$ for all $n > N$.

Therefore, combining the last paragraph with Lemma 12, extracting subsequences and relabeling we can assume:

There is $\varepsilon_n \to 0$ and $p_n \in C_1^+(K)$ such that, letting

$$D_n = \left\{ t \in I : p_n \boldsymbol{a}_n(t) \geq p_n \boldsymbol{\omega}_n(t) + \frac{1}{n} \right\},$$

one has $\lambda(C_n) \leq 1/2^n$, $\lambda(D_n) \leq 1/2^n$, and $p_n a \leq p_n \boldsymbol{\omega}_n(t) - 2\varepsilon_n$ implies

$$\boldsymbol{a}_n(t) + \varepsilon_n b \succsim_t a \quad \text{for a.e.} \quad t \in I/C_n.$$

Lemma 13. *Suppose there are $K'_n, K''_n \in \mathscr{C}(K)$ and $A'_n, A''_n, A'''_n \subset I$ such that:*

(i) $\varepsilon_n, K'_n, K''_n, A'_n, A''_n, \boldsymbol{a}_n, p_n$ *satisfy the conditions of Lemma 9;*
(ii) $A''_n \subset A'''_n$; $A'''_{n+1} \subset A'''_n$ *and* $\lambda(A'''_n) \to 0$;
(iii) *for a.e.* $t \in I/A'''_n$, $\boldsymbol{a}_n(t), \boldsymbol{\omega}_n(t) \in \Omega_{K_n''}$.

Then ***a*** *induces a core distribution.*

Proof. By notational convenience let $p_n^* = p_n \mid K''_n$, so

$$p_n^* \in \mathscr{C}_1^+(K''_n).$$

By Lemmas 9 and 8 we can assume $(K''_n, p^*_n) \to (K, p)$ for some $p \in C^+_1(K)$. Pick a fixed N; then, for a.e. $t \in I/A'''_n$ if $n > N$, one has

$$\boldsymbol{a}_n(t), \boldsymbol{\omega}_n(t) \in K''_n,$$

$$p_n\boldsymbol{a}_n(t) = p^*_n\boldsymbol{a}_n(t),$$

$$p_n\boldsymbol{\omega}_n(t) = p^*_n\boldsymbol{\omega}_n(t),$$

and

$$|p_n\boldsymbol{a}_n(t) - p_n\boldsymbol{\omega}_n(t)| \leq \min\{4\varepsilon_n, 1/n\}.$$

Hence, by Lemma 4 [remember $\boldsymbol{a}_n(t) \to \boldsymbol{a}(t)$ a.e.], $p_n\boldsymbol{a}_n(t) \to p\boldsymbol{a}(t)$ and $p_n\boldsymbol{\omega}_n(t) \to p\boldsymbol{\omega}(t)$ for a.e. $t \in I/A'''_n$ or, simply, since N is arbitrary and $\lambda(A'''_n) \to 0$, for a.e. $t \in I$. So, $p\boldsymbol{a}(t) = p\boldsymbol{\omega}(t)$ for a.e. $t \in I$.

Fix again an N. For a.e. $t \in I/A'''_n$ if $a \succ_t \boldsymbol{a}(t)$, then for $\mu > 0$ small enough and $n > N'$ ($N' > N$ large enough) there is (use Fact 6) $\boldsymbol{a}_n \in \Omega_{K_n''}$ such that

$$a_n \to a \quad \text{and} \quad a_n - \mu b \succ_t \boldsymbol{a}_n(t) + \varepsilon_n b.$$

Therefore,

$$p^*_n a_n - \mu b = p_n a_n - \mu b > p_n\boldsymbol{\omega}_n(t) - 2\varepsilon_n = p^*_n\boldsymbol{\omega}_n(t) - 2\varepsilon_n.$$

Hence (by Lemma 4) $pa - \mu b \geq p\boldsymbol{\omega}(t)$, i.e., $pa > p\boldsymbol{\omega}(t)$. Since N is arbitrary and $\lambda(A'''_n) \to 0$, we conclude that for a.e. $t \in I$, $a \succ_t \boldsymbol{a}(t)$ implies $pa > p\boldsymbol{\omega}(t)$. ■

Therefore, to conclude the proof of Theorem 2 we need to show the existence of A'_n, K'_n, A''_n, K''_n, A'''_n with the required properties. This is done, respectively, in the next five paragraphs; we skip some (easy) verification details.

(1) *Construction of A'_n*: Let $A'_n = \bigcup_{k>n} (C_k \cup D_k)$. Then $\lambda(A'_n) \leq \frac{1}{2^{n-2}}$.

(2) *Construction of K'_n*: For every n, let $V_n \subset K^d$ be the union of the B^n_i, $1 < i \leq m_n$, for which

$$\left(\textstyle\int_I \boldsymbol{\omega}\right)\left(B^n_i\right) = \left(\textstyle\int_{A_n'} \boldsymbol{a}\right)\left(B^n_i\right).$$

Take

$$V'_n = \bigcup_{k>n} V_k$$

and put

$$K_n' = K_n / V_n'.$$

It is simply checked that A_n' and K_n' satisfy the properties required by Lemma 9.

Observe that

$$(\textstyle\int_I \boldsymbol{\omega})(V_n) = (\textstyle\int_{A_{n'}} \boldsymbol{a})(V_n) \leq \frac{\alpha}{2^{n-2}}$$

and so,

$$(\textstyle\int_I \boldsymbol{\omega})(V_n') \leq \frac{\alpha}{2^{n-3}}.$$

(3) *Construction of A_n''*: Denote

$$F_n = \{t \in I : \boldsymbol{\omega}(t)(V_n') > 0\} = \{t \in I : \boldsymbol{\omega}(t)(V_n') \geq 1\};$$

then

$$\lambda(F_n) \leq \alpha \frac{1}{2^{n-3}}.$$

Let

$$F_n' = \bigcup_{k>n} F_k;$$

so,

$$\lambda(F_n') \leq \frac{\alpha}{2^{n-4}}.$$

Put

$$A_n'' = A_n' \cup F_n';$$

then

$$\lambda(A_n'') \leq \frac{\alpha}{2^{n-5}}$$

and it is easily checked that K_n' and A_n'' satisfy the properties required by Lemma 9.

(4) *Construction of K_n''*: We repeat Step (2). Let O_n be the union of the B_i^n, $1 < i \leq m_n$, for which

$$\left(\textstyle\int_I \boldsymbol{\omega}\right)\left(B_i^n\right) = \left(\textstyle\int_{A_{n''}} \boldsymbol{a}\right)\left(B_i^n\right).$$

Take

$$O'_n = \bigcup_{k>n} O_k$$

and put

$$K''_n = K_n / O'_n;$$

note that, indeed, $K''_n \subset K'_n$. Now we have

$$\left(\textstyle\int_I \boldsymbol{\omega}\right)\left(O'_n\right) \leq \frac{\alpha^2}{2^{n-6}}.$$

(5) *Construction of* A''_n: We repeat Step (3). Note for the moment that for a.e. $t \in I/A''_n$, $\boldsymbol{a}_n(t) \in \Omega_{K_{n''}}$. Denote

$$G_n = \{t \in I : \boldsymbol{\omega}(t)(O'_n) > 0\},$$

and let

$$G'_n = \bigcup_{k>n} G_k.$$

Then

$$\lambda(G_n) \leq \frac{\alpha^2}{2^{n-6}},$$

and so,

$$\lambda(G'_n) \leq \frac{\alpha^2}{2^{n-7}},$$

i.e., $\lambda(G'_n) \to 0$. Put $A'''_n = A''_n \cup G'_n$.

This concludes the proof of Theorem 2.

Appendix

We give an example to show the need of the compactness of agents' characteristics assumption.

Let $K = \{0, 1, \frac{1}{2}, \cdots, \frac{1}{n}, \cdots\} \cup \{h\}$. A consumer of type n ($\succsim_n$, ω_n) is defined as follows:

$$\boldsymbol{\omega}_n\left(K^d \Big/ \left\{\frac{1}{n}\right\}\right) = 0, \quad \boldsymbol{\omega}_n\left(\frac{1}{n}\right) = 1, \quad \boldsymbol{\omega}_n(h) = 1;$$

a utility function u_n for $\succsim_n$ is given by

$$u_n(a) = (n+1)a(h) \quad \text{if} \quad a\left(\frac{1}{n+1}\right) \geq 1$$

and

$$u_n(a) = a(h) \quad \text{if} \quad a\left(\frac{1}{n+1}\right) = 0.$$

Take an economy ν giving weight $1/2^n$ to traders of type n. Clearly,

$$\operatorname{supp}\left(\int_\Omega i \, \mathrm{d}\nu_E\right) = K.$$

Denote

$$J_n = \left\{0, \frac{1}{n}, \frac{1}{n+1}, \cdots\right\}, \qquad S_n = \left(\int_\Omega i \, \mathrm{d}\nu_E\right)(J_n);$$

then

$$S_n = \frac{1}{2^{n-1}}.$$

Let $p \in C_1^+(K)$ be an equilibrium price and pick N such that

$$\frac{1}{2}(N+1) > 2 + p(0)$$

and

$$\left|p\left(\frac{1}{n}\right) - p(0)\right| < \frac{1}{4} \quad \text{for} \quad n \geq N.$$

Consider a trader of type $n \geq N$: his wealth is $1 + p(1/n)$; if he buys nothing of commodity $1/(n+1)$, then his utility is $1 + p(1/n)$, if he buys a unit his utility is

$$(n+1)\left(1 + p\left(\frac{1}{n}\right) - p\left(\frac{1}{n+1}\right)\right) > 1 + p\left(\frac{1}{n}\right);$$

hence every consumer of type $n \geq N$ demands a unit of commodity $1/(n+1)$; but then, if τ is an equilibrium distribution and

$$D_n = \left(\int_\Omega i \, \mathrm{d}\tau_\Omega\right)(J_n),$$

we have

$$D_{N+1} = \frac{1}{2^{N-1}} > \frac{1}{2^N} = S_{N+1},$$

i.e., demand is greater than supply. Hence, no equilibrium price system exists.

The essence of this example is the lack of compactness in consumers' preferences (as $N \to \infty$, $\succsim_n$ has no limit point in $\mathscr{P}$); commodities are not uniformly substitutable and as a consequence, no finite price for the commodity {0} will clear the market.

We should remark that the preferences of this example do not satisfy Condition VIII(iii) either. However, we already pointed out that, if, as in the example, consumers are endowed with no more than one unit of differentiated commodities, then Condition VIII(iii) is dispensable in Theorem 1, so the existence failure has to be attributed to the lack of preference compactness.

References

Arrow, K. and F. Hahn, 1971, General competitive analysis (Holden-Day, San Francisco).

Aumann, R., 1964, Markets with a continuum of traders, Econometrica.

Aumann, R., 1966, Existence of competitive equilibrium in markets with a continuum of traders, Econometrica.

Bewley, T., 1970, Equilibrium theory with an infinite-dimensional commodity space, Ph.D. dissertation (University of California, Berkeley).

Bewley, T., 1972, Existence of equilibria in economies with infinitely many commodities, Journal of Economic Theory.

Bewley, T., 1973, The equality of the core and the set of equilibria in economies with infinitely many commodities and a continuum of agents, International Economic Review.

Broome, J., 1972, Approximate equilibrium in economies with indivisible commodities, Journal of Economic Theory.

Chamberlin, E. H., 1956, The theory of monopolistic competition, 7th ed. (Harvard University Press, Cambridge).

Debreu, G., 1959, Theory of value (Wiley, New York).

Debreu, G. and H. Scarf, 1963, A limit theorem on the core of an economy, International Economic Review.

Dierker, E., 1971, Equilibrium analysis of exchange economies with indivisible commodities, Econometrica.

Gabszewicz, J., 1968, Coeurs et allocations concurrentielles dans des économies d'échange avec un continu de biens, Thesis (Louvain, Belgium).

Gabszewicz, J. and J. Mertens, 1971, An equilibrium theorem for the cores of an economy whose atoms are not 'too' big, Econometrica.

Grodal, B., 1974, A note on the space of preference relations, Journal of Mathematical Economics.

Hart, S., W. Hildenbrand and E. Kohlberg, 1974, On equilibrium allocations as distributions on the commodity space, Journal of Mathematical Economics.

Henry, C., 1970, Indivisibilités dans une économie d'échange, Econometrica.

Hildenbrand, W., 1974, Core and equilibria of a large economy (Princeton University Press, Princeton).

Houthakker, H., 1952, Compensated changes in quantities and qualities consumed, Review of Economic Studies.

Kluvánek, I., 1973, The range of a vector-valued measure, Mathematical Systems Theory.

Lancaster, K., 1966, A new approach to consumer theory, Journal of Political Economy.

Mas-Colell, A., 1974, An equilibrium existence theorem without complete or transitive preferences, Journal of Mathematical Economics.

Neveu, J., 1965, Mathematical foundations of the calculus of probability (Holden-Day, San Francisco).

Ostroy, J., 1973, Representations of large economies: The equivalence theorem, mimeo. (University of California, Los Angeles).

Robinson, J., 1933, The economics of imperfect competition (Macmillan, London).

Rosen, S., 1974, Hedonic prices and implicit markets: Product differentiation in pure competition, Journal of Political Economy.

Royden, H., 1968, Real analysis, 2nd ed. (Macmillan, New York).

Samuelson, P., 1969, The monopolistic competition revolution, in: R. Kuenne, ed., Monopolistic competition theory: Studies in impact (Wiley, New York).

Schaefer, H., 1971, Topological vector spaces (Springer, New York).

Schmeidler, D., 1969, Competitive equilibria in markets with a continuum of traders and incomplete preferences, Econometrica.

— 3 —

On the Equilibrium Price Set of an Exchange Economy*

1. Introduction

Consider a pure exchange economy composed of a finite number of traders with strictly positive initial endowments and continuous, monotone, strictly convex preferences. It is well known that the intersection of the unit sphere with the set of price equilibria for such an economy is nonempty and compact. Under smoothness hypotheses on preferences and generic (i.e., non-degeneracy) conditions further (fixed-point index type) strong restrictions on the equilibrium price set can be derived; this was done by Dierker (1972). In this paper we provide converses to the above statements, i.e., we prove that, imprecisely speaking, if the number of commodities is greater than two, then every pattern of equilibria compatible with the above referred to properties can arise from an economy in the class we consider.

It is obvious that this characterization of the equilibrium price set problem [already studied by Sonnenschein (1972)] is closely related to the problem of characterizing excess demand functions defined on compact sets of prices. In fact, the solution to the first problem has had to wait for a solution to the second to be found; this has been accomplished only recently; see Sonnenschein (1973), Mantel (1974, 1976), Debreu (1974). Our proof here amounts to a refinement of the one by Debreu (1974) leading to a sharper version of his result, sharp enough for the purposes of this paper.

Journal of Mathematical Economics 4 (1977), 117–126.

* I am very indebted to G. Debreu, L. Shapley, and R. Mantel for their comments. Financial support from NSF grants SOC72-05551A02 and SOC73-05650A01 is gratefully acknowledged.

2. Statement of Results

The commodity space is R^l; $R^l_+ = \{p \in R^l : p \geqq 0\}$, $P = \{p \in R^l : p \gg 0\}$, $\hat{P} = R^l_+ \backslash \{0\}$, $S = \{p \in P : ||p|| = 1\}$, $S_\varepsilon = \{p \in S : p^i \geqq \varepsilon$ *for all* $i\}$; *for* $p \in R^l$, $T_p = \{x \in R^l : px = 0\}$; $e = (1, \ldots, 1) \in R^l$.[1]

In our context an *excess demand function* $f: S \to R^l$ is a continuous function satisfying:

(W) $pf(p) = 0$ for every $p \in S$,
(BB) there is $k \in R$ such that for every $p \in S, f(p) > ke$,
(BC) if $p_n \to p \in \partial S$, $p_n \in S$, then $\lim n\, ||f(p_n)|| = \infty$.

Note that the sum of two excess demand functions is an excess demand function.

For every excess demand function f let the *set of equilibria* be $E_f = \{p \in S : f(p) = 0\}$; under the assumptions made E_f is non-empty and compact [see Debreu (1970)].

All the economies to be considered will be of pure exchange and of the form $\mathscr{E} = \{(\succsim_i, \omega_i, R^l_+)\}^l_{i=1}$, where R^l_+ is the consumption set, $\succsim_i$ is a continuous, monotone, strictly convex preference relation on R^l_+, and $\omega_i \in P$. To avoid repetition, from now on an *economy* is defined to have those properties (hence, economies always have l participants).

Given an economy $\mathscr{E}$, an (aggregate) excess demand function f corresponding to $\mathscr{E}$ (we also say that $\mathscr{E}$ generates f) is defined in the usual manner, i.e.,

$$f(p) = \sum_{t=1}^{l} \{x \in R^t_+ \mid px \leqq p\omega_i \quad \text{and} \quad py \leqq p\omega_i,\, y \geqq 0,$$

$$\text{implies} \quad x \succsim_i y''\} - \sum_{i=l}^{l} \{\omega_i\}.$$

It is known that f will in fact be an excess demand function, i.e., satisfy (W), (BB), and (BC)—see Arrow and Hahn (1971, ch. 4, theorem 8).

It is not known if every excess demand function can be generated by an economy. Following the work of Sonnenschein (1972, 1973), and Mantel (1974), Debreu (1974) showed that for any excess demand f and $\varepsilon > 0$

1 $||\ ||$ denotes the Euclidean norm. Subscripts indicate vectors and superscripts components of vectors. The boundary, interior, and closure of a set A are denoted, respectively, by ∂A, Int A, and $\bar{A}$. Neighborhood always mean open neighborhood.

there is an excess demand f^* such that $f^* |S_\varepsilon = f|S_\varepsilon$ and f^* is generated by an economy [of course, Debreu does not impose conditions (BB) and (BC)]. There is no presumption, however, to the effect that $E_{f^*} \subset E_f$, which makes the result fall just short of what we need for the characterization of the equilibrium price set problem. The main result of this paper is an extension of Debreu's theorem which renders it readily usuable for our purposes.

Theorem. *If f: $S \to R^l$ is an excess demand function and* $\varepsilon > 0$, *then there is* $\mu < \varepsilon$ *and an excess demand function* f^* *such that* $f^*|S_\mu = f|S_\mu$, $E_{f^*} = E_f \subset S_\mu$ *and* f^* *is generated by an economy.*

The following corollary is immediate:

Corollary 1. *Let $A \in S$ be a non-empty, compact set; then there is an excess demand f generated by an economy $\mathscr{E}$ such that* $A = F_f$.

Proof. Pick some $z \in A$; for every $p \in S$ let $h(p)$ be the projection of $z - p$ on T_p. Define $\alpha(p) = \min \{||p - q|| : q \in A\}$; $p + \gamma(p)h(p) \in \partial P$. The function $\gamma(p)$ is continuous on S since $S \subset P$, P is open, convex, and h is continuous. Define $f(p) = \alpha(p)\gamma(p)h(p)$. It is easily checked that f is an excess demand function and $f(p) = 0$ if and only if $p \in A$. Apply, then, the theorem. ∎

As a characterization of the equilibrium price set result, Corollary 1 is not yet very satisfactory; on the one hand, many of the equilibria in $A = E_f$ will typically be 'degenerate'; on the other hand, plenty of relevant information about the nature of an equilibrium price vector is contained not only in its position in S but also in the form of the linear derivative map of the excess demand function at this price vector.[2]

Let f be a C^1 excess demand function. At every $p \in E_f$ the map $Df(p)$: $T_p \to R^l$ carried T_p into itself. So, from now on we shall regard $Df(p)$ as a map from T_p into T_p (if $p \in E_f$!). If f satisfies

(R) $$|Df(p)| \neq 0 \quad \text{for every} \quad p \in E_f,^3$$

2. In contrast, the properties of the derivative map of excess demand at non-equilibrium prices are very sensitive to such irrelevant matters as normalization.

3. $|Df(p)|$ is the determinant of $Df(p)$; sign $|Df(p)| = +1, 0, -1$, according to if $|Df(p)| > 0, = 0, < 0$; f is C^1 on a closed set K if it can be extended to a C^1 function in a neighborhood of K.

then E_f is finite, and Dierker (1972) has shown that

$$\sum_{p \in E_f} \text{sign } |Df(p)| = (-1)^{l+1}.$$

It follows from results of Debreu (1970) that condition (R) is generic (i.e., it holds for 'almost every' economy, where 'almost every' is appropriately defined) in the class of economies having C^1 demand functions.

The next corollary shows, in a particularly strong form, that (at least, if the number of commodities is greater than two) the fixed-point-index condition of Dierker exhausts the restrictions on equilibrium prices derivable from the underlying economic model of this paper.

Corollary 2. *Suppose we are given a compact set $K \subset S$ and a function f: $K \to R^l$ such that, letting $E = \{p \in K : f(p) = 0\}$: (i) K is an $l - 1$ smooth, compact manifold (for example, K may be a union of a finite collection of pairwise disjoint closed discs); (ii) $S \backslash K$ is connected; (iii) f is C^l; (iv) $pf(p) = 0$ for every $p \in K$; (v) $|Df(p)| \neq 0$ for $p \in E$; and (vi) $E \neq \phi$, $E \subset \text{Int } K$, and $\Sigma_{p \in E}$ sign $|Df(p)| = (-1)^{l+1}$. Then there is an excess demand function f^* such that f^* is generated by an economy $f^* | K = f$ and $f^*(p) = 0$ only if $p \in K$.*

Proof. Let $B \subset S$ be a $(l - 1)$ ball (up to diffeomorphism) which contains in its interior $K \cup \{(1/l^{\frac{1}{2}})e\}$. Define $g : \partial B \to R^l$ by $g(p) = e - (pe)p$. Parametrizing B by the first $(l - 1)$ coordinates, we can regard B as an open subset of R^{l-1}; analogously, we replace $f(p)$, $g(p)$ by their projections on the first $l - 1$ coordinates, which, abusing notation, we denote by the same symbol. Let $W = \overline{B \backslash K}$; then W is a compact, connected, oriented $(l - 1)$ manifold and $\partial W = \partial B \cup \partial K$. Define $F : \partial W \to S^{l-2}$ by letting $F(p)$ equal $f(p)/\|f(p)\|$ if $p \in \partial K$ and $g(p)/\|g(p)\|$ if $p \in \partial B$. Clearly, the proof is finished if the map F can be extended from ∂W to W or equivalently [this is Hopf's theorem; see Guillemin and Pollack (1974, p. 145)] if the degree of F is zero. But

$$\begin{aligned}\deg F &= -\deg f|\partial K + \deg g|\partial B \\ &= -\sum_{p \in E} \text{sign } |Df(p)| + \deg g|\partial B \\ &= -(-1)^{l+1} + (-1)^{l+1} = 0. \quad \blacksquare\end{aligned}$$

No attempt has been made to state the most general possible version of the corollary. The requirement that $S \backslash K$ be connected has strong implications only for $l = 2$. It is clear in this case that if we represent prices in a line, then the equilibrium prices associated with positively sloped excess demand should alternate with the negatively sloped ones.

3. Proof of the Theorem

(1) For every $p \in S$ let $g_e(p) = e - (pe)p$. The fact proved in the following lemma is well known:

Lemma 1. If f is an excess demand function there is $\delta > 0$ such that $f(p)g_e(p) > 0$ whenever $p \in S \backslash S_\delta$.

Proof. By (BB) and (BC) there is $\delta > 0$ such that if $p \in S \backslash S_\delta$, then

$$\sum_{i=1}^{l} f^i(p) > l.$$

But then

$$f(p)g_e(p) = \sum_{i=1}^{l} f^i(p) > 0. \qquad \blacksquare$$

In view of Lemma 1 the theorem is an obvious corollary of the next proposition; its statement is somewhat technical but it yields results which are important in applications [see Mas-Colell (1977)]; among other things it makes clear that the norm of the initial endowments of the economy to be constructed only depends on the norm of the values of excess demand (and not, for example, on the norm of derivatives were they to exist).

Proposition. For every $\varepsilon > 0$ there is $0 < \mu < \varepsilon$ and a function $k : (0, \infty)$ such that if f: $S \to R^l$ is a continuous function satisfying $pf(p) = 0$ for every $p \in S$ and $f(p)g_e(p) > 0$ for $p \in S \backslash S_\varepsilon$ then there exists an excess demand function f^* satisfying: (i) $f^*|S_\varepsilon = f|S_\varepsilon$, (ii) $f^*(p)g_e(p) > 0$ for $p \in S \backslash S_\varepsilon$, (iii) f^* is generated by an economy $\{(R_+, \succsim_i, \omega_i)\}_{i=1}^{l}$ with, for all i, $||\omega_i|| \leqq l(2 + k(r))$ for any $r \geqq \text{Max}_{p \in S_\mu} ||f(p)||$, and $\succsim_i$ continuous, monotone, strictly convex.

Notice that in the proposition μ and k depend only on ε. No attempt is made to obtain a sharp bound for $||\omega_i||$.

(2) We proceed now to prove the proposition:

We state first some properties of functions from P, or S, to R.

(I) $v : P \to R$ is proper, strictly convex, and C^2. Moreover, $p_n \to p \in \partial P$ implies $\|\nabla v(p_n)\| \to \infty$ and there is an $r > 0$ such that if $\|p\| \leqq 1$, then $p\nabla v(p) \leqq r$ and $\nabla v(p) \leqq 0$.

(II) $v : S \to R$ is the restriction to S of a function on P satisfying (I); see fig. 1.

Note that the sum of functions satisfying (I) [resp. (II)] satisfies (I) [resp. (II)]. A function is proper if inverse images of compact sets are compact.

Let $v : S \to R$ satisfy (II); then v is minimized in at most one point and for every $r \in v(S)$ the cone spanned by $v^{-1}((-\infty, r])$ is strictly convex [indeed, let v be defined on P and satisfy (I); if $v(p), v(q) \leqq r$, $p, q \in S$, $p \neq q$, then, for $0 < t < 1$, $\|tp + (1 - t)q\| < 1$ and $v(tp + (1 - t)q) < r$; since v is decreasing on $\{q' : \|q'\| < 1\}$, $v((1/\|tp + (1 - t)q\|)(tp + (1 - t)q)) < r$]. Therefore, if v satisfies (II), then for all $p \in S$, $\nabla v(p) \in T_p$ is normal to a supporting hyperplane at p of the cone spanned by $v^{-1}((-\infty, v(p)])$; hence, $v(p) > v(q)$ implies $\nabla v(p)(q - (pq)p) < 0$. See fig. 1.

Observe that if $v: S \to R$ satisfies (II), then $p_n \to p \in \partial S$ implies $\|\nabla v(p_n)\| \to \infty$ because, letting $v = v' \mid S$ for a $v': P \to R$ satisfying (I), $\nabla v(p_n) = \nabla v'(p_n) - (p_n \nabla v'(p_n))p_n$.

Lemma 2. *Let* $v : S \to R$ *satisfy* (II). *Suppose that* $f : S \to R^l$ *is an excess demand function such that, for all* $p \in S$, $f(p) > \alpha e$ *(where* $\alpha > -\infty$*) and* $f(p) = -\beta(p)\, \nabla v(p)$ *where* $\beta(p)$ *is positive and uniformly bounded away from* 0. *Then* f *is generated by a* $(\succsim, \omega, R^l_+)$ *where* $\succsim$ *is a continuous, monotone, strictly convex preference relation on* R^l_+ *and* $\omega \in P$, $\|\omega\| \leqq l\,|\alpha|$.

Proof. A careful reading of Debreu's proof (1974) would validate Lemma 2. For the sake of completeness, and skipping details, we outline here a somewhat different demonstration.

Obviously, we can take $\omega = 0$ and look for a $\succsim$ defined on R^l. We will show only the existence of a continuous and monotone $\succsim$ generating f; $\succsim$ could then be 'convexified' by using standard methods [see Hildenbrand (1975, 1.1, problem 7)]; in order to have $\succsim$ strictly convex one would have to apply, in a second step, Debreu's (1974) technique.

There is a unique $\bar{p} \in S$ for which $f(\bar{p}) = 0$. Divide R^l in three regions: $\hat{P}$,

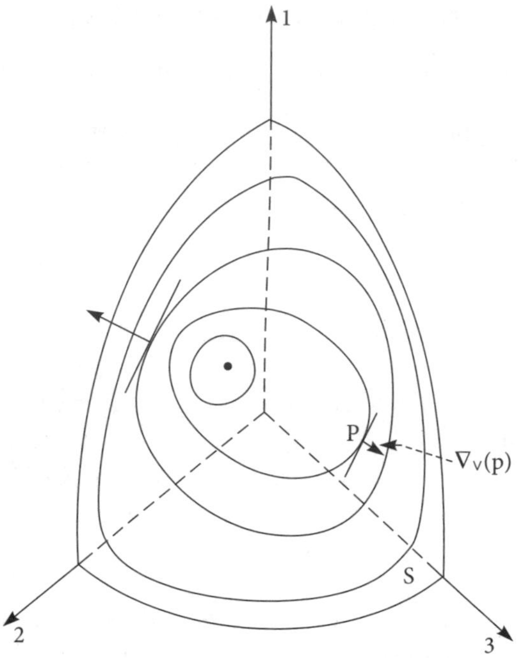

Figure 1

$A = \{x \mid x \notin \hat{P}, \bar{p}x > 0\}$, $B = \{x \mid \bar{p}x \leqq 0\}$. Let $u : S \to [0, 1]$ be a continuous function such that $u(\bar{p}) = 0$ and $u(p) \geqq u(q)$ if and only if $v(p) \geqq v(q)$.

By the properties of v there is for every $x \in A$ a unique $p(x) \in S$ such that x belongs to the positive ray spanned by $f(p(x))$. The function $x \mapsto p(x)$ is continuous. Let $x_n \in A$, $x_n \to x$, $x \notin A$, then $p(x_n) \to \partial S$ if $x \in \hat{P}$ or $p(x_n) \to \bar{p}$ if $x \in B$. One also sees that $p \in S$, $x \in A$, $px \leqq 0$, and $x \notin \{tf(p) : t \geqq 0\}$ implies $u(p(x)) < u(p)$ (see fig. 2).

Define a function $\xi : R^l \to R$ as follows: for $x \in A$ write $x = tf(p(x))$ and take $\xi(x) = tu(p(x))$ if $t \leqq 1$ or $\xi(x) = u(p(x)) - ||x - f(p(x))||$ if $t > 1$; for $x \in B$ take $\xi(x) = -||x||$; for $x \in \hat{P}$ take $\xi(x) = \min x^i$.

The function ξ satisfies: (i) ξ is continuous [this is so because u is bounded, $u(\bar{p}) = 0$, f is continuous, and $\lim ||f(p_n)|| \to \infty$ as $p_n \to \partial S$]; (ii) ξ has no local maxima; (iii) for any $p \in S$, if $px \leqq 0$ and $x \neq f(p)$, then $\xi(x) < \xi(f(p))$. Let $\hat{\xi}(x) = \max \{\xi(y): y \leqq x\}$; the function $x \mapsto \hat{\xi}(x)$ is well defined and, by (i) above, continuous. By (ii), $x \gg y$ implies $\hat{\xi}(x) > \hat{\xi}(y)$. It is, moreover, not difficult to check that, for all $p \in S$, $px \leqq 0$ and $x \neq f(p)$

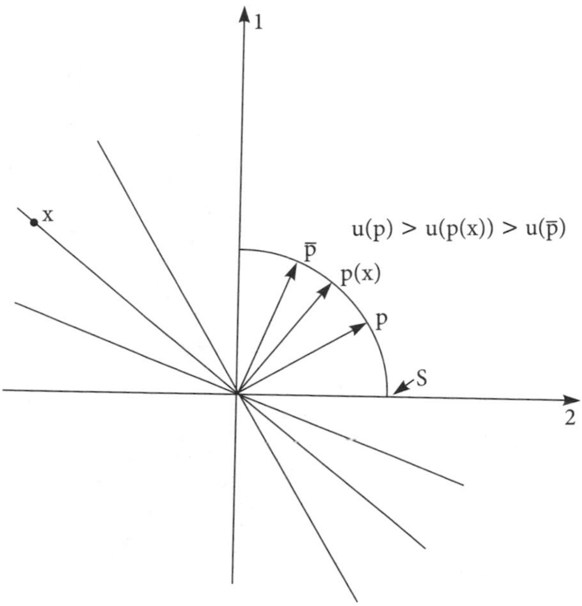

Figure 2

implies $\hat{\xi}(x) > \hat{\xi}(f(p))$. Hence the preference relation given by $x \succsim y$ if and only if $\hat{\xi}(x) > \hat{\xi}(y)$ is continuous, monotone, and generates f. ■

For any $a \in P$ define $u_a : S \to R$ by $u_a(p) = (1/2\|a\|)\ \|p - a\|^2$; *if* $a \geqq e$, u_u satisfies (II). Observe that $-\nabla u_a(p) = (1/\|a\|)\ (a - (pa)p)$; see fig. 3.

Lemma 3. For any $a \in P$ with $a \geqq e$ and compact K with $(1/\|a\|)\ a \in K$, *there is a function $v : S \to R$ satisfying* (II) *and such that: (i) $v \mid K = u_a$ and $v(p) \geqq u_a(p)$ for all p; (ii) $\nabla v(p) \leqq 2e$ for all p; see fig. 4.*

Proof. Define $\xi : P \to R$ by $\xi(p) = -(1/l)(\sum_{i=1}^{l} \log p^i)$ and $\hat{u}_a : P \to R$ by $\hat{u}_a(p) = (1/2\|a\|)\ \|p - a\|^2$ (so, $u_a = \hat{u}_a \mid S$). Both ξ and $\hat{u}_a$ satisfy (I). Clearly, for $p \in S$, $\nabla(\xi \mid S)(p) = p - (1/l)(1/p^1, \ldots, 1/p^l) < 2e$.

Let $B \subset S$ be a neighborhood of K with $\overline{B} \subset S$ and define $\gamma : P \to R$ by $\gamma(p) = \max\ \{0, \xi(p) - s\}$ where s is chosen large enough to have $\gamma(p) = 0$ on a neighborhood of B; γ is convex and we can smooth it out around $\gamma^{-1}(s)$ so as to obtain a convex function $\hat{\gamma}$; $P \to S$ which satisfies (I)—except for strict convexity—and $\hat{\gamma}(p) \geqq 0$, $\nabla(\hat{\gamma} \mid S)(p) \leqq e$ for $p \in S$, $\hat{\gamma}(p) = 0$ for

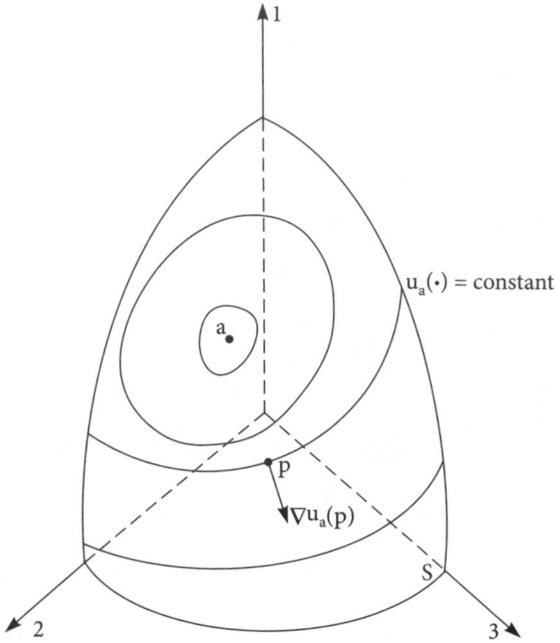

Figure 3

$p \in B$. Take then $v = \hat{u}_a + \hat{\gamma} \mid S = u_a + \hat{\gamma} \mid S$; it is clear that v satisfies (II) and $\nabla v(p) = (1/\|a\|)\,((pa)p - a) + \nabla(\hat{\gamma} \mid S)(p) \leqq 2e$. ■

Let $\varepsilon > 0$. This ε will remain fixed for the rest of the proof. For a $\delta > 0$ choose $a_i \in S$ such that $a_i^j \geqq e$ and $(1/\|a^i\|)\, a_i^j = \delta$ for every $j \neq i$; δ is chosen small enough to guarantee that $S_{\varepsilon/2}$ is contained in the interior of the convex cone $\Gamma \subset R^l$ spanned by $\{a_1, \ldots, a_l\}$. Choose $\mu > 0$ such that $\Gamma < S \subset S\mu$.

For any $r > 0$ let $B_r \subset R^l$ be the closed r ball. Given $r > 0$ pick $\theta > 0$ such that, for every $x \in B_r$ and $p \in S_{\varepsilon/2}$, $x + \theta p \in \Gamma$. Such θ does obviously exist (remember $S_{\varepsilon/2} \subset \operatorname{Int} \Gamma$). We put $k(r) = \theta + r$.

For every i pick a function $v_i : S \to R$ satisfying the conclusions of Lemma 3 with respect to a_i and $\Gamma \cap S$. The function $\sum_{i=1}^{l} u_{a_i}$ attains its minimum value at the unique point $p = (1/l^{\frac{1}{2}})\, e \in S$. Since $v_i \geqq u_{ai}$ and $v_i\, |S_\varepsilon = u_{ai}|\, S_\varepsilon$ we get $\sum_{i=1}^{l} v_i(p) > \sum_{i=1}^{l} v_i((1/l^{\frac{1}{2}})e)$ for all $p \neq (1/l^{\frac{1}{2}})e$; since $\sum_{i=1}^{l} v_i$ satisfies (II), we conclude that $\sum_{i=1}^{l} \nabla v_i(p)(e - (pe)p) = \sum_{i=1}^{l} \nabla v_i(p) g_e(p) < 0$ for all $p \neq (1/l^{\frac{1}{2}})e$.

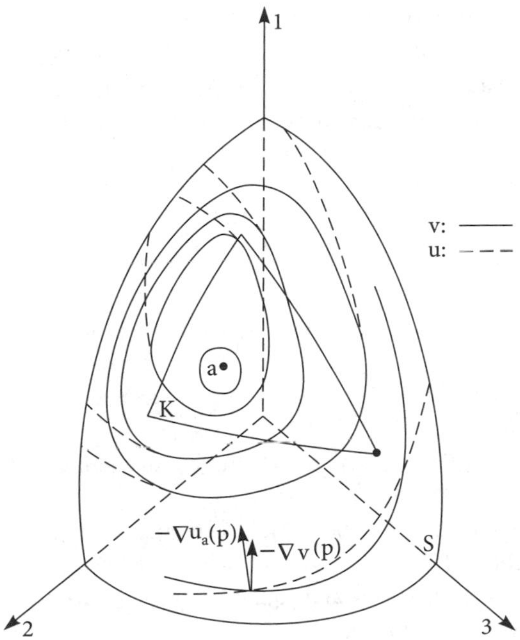

Figure 4

Let f be an excess demand function satisfying $f(p)g_e(p) > 0$ for $p \in S\backslash S_\varepsilon$. To obtain a f^* fulfilling the conclusions of the proposition we proceed as in Debreu (1974).

Pick $\bar{r} > \text{Max}_{p \in S_\mu} \|f(p)\|$. Then, for every $p \in S_{\varepsilon/2}$, $f(p) + (k(\bar{r}) - \bar{r})p \in \Gamma$. Since the vectors $\{a_1, \ldots, a_l\}$ are linearly independent we can write, for $P \in S_{\varepsilon/2}$, $f(p) + (k(\bar{r}) - \bar{r})p = \sum_{i=1}^{l} \beta_i(p)a_i$, $\beta_i(p) > 0$ in a unique and continuous fashion; moreover, for all i, $\|\beta_i(p)a_i\| \leqq \|f(p) + (k(\bar{r}) - \bar{r})p\| \leqq k(\bar{r})$. Observe that $-\nabla u_{a_i}(p) = a_i - (pa_i)p$ is nothing but the perpendicular projection of a_i on T_p. So, projecting perpendicularly into T_p we get, for $p \in S_{\varepsilon/2}$, $f(p) = \sum_{i=1}^{l} \beta_i(p)(-\nabla v_i(p))$ and, of course, $\|\beta_i(p)\nabla v_i(p)\| \leqq \|\beta_i(p)a_i\| \leqq k(\bar{r})$ for all i.

Let $\eta : S \to [0, 1]$ be C^2 and such that $\eta(p) = 1$ for $p \in S\backslash S_{\varepsilon/2}$ and $\eta(p) = 0$ for $p \in S_\varepsilon$. Define $f_i : S \to R^l$ by $f_i(p) = -(\eta(p)) + (1 - \eta(p))\beta_i(p))\nabla v_i(p)$. Let $f^* = \sum_{i=1}^{l} f_i$ and note that every f_i, and so f^*, are excess demand functions.

By Lemmas 2 and 3 every f_i can be generated by a $(\succsim, \omega, R_+^l)$ with $\succsim$ continuous, monotone, strictly convex and with $\|\omega\| \leqq l \max\{k(\bar{r}), 2\} \leqq$

$l(k(s(\overline{r})) + 2)$. By construction $f^*(p) = f(p)$ for $p \in S_\varepsilon$. If $p \in S \backslash S_\varepsilon$, then either $p \in S \backslash S_{\varepsilon/2}$ in which case $-f^*(p)g_e(p) = \sum_{i=1}^{l} \nabla v_i(p)g_e(p) < 0$, or $p \in S_{\varepsilon/} \backslash S_{\varepsilon 2}$ in which case we also have $f^*(p)g_e(p) = -\eta(p)(\sum_{i=1}^{l} \nabla v_i(p)g_e(p)) + (1 - \eta(p))f(p)g_e(p) > 0$ because both terms of the sum are positive. Q.E.D.

References

Arrow, Kenneth and Frank Hahn, 1971, General competitive analysis (Holden–Day, San Francisco, CA).

Debreu, Gerard, 1970, Economies with a finite set of equilibria, Econometrica 38, 387–392.

Debreu, Gerard, 1974, Excess demand functions, Journal of Mathematical Economics 1, 15–23.

Dierker, Egbert, 1972, Two remarks on the number of equilibria of an economy, Econometrica 40, 951–955.

Guillemin, Victor and Alan Pollack, 1974, Differential topology (Prentice-Hall, Englewood Cliffs, NJ).

Hildenbrand, Werner, 1975, Core and equilibria of a large economy (Princeton University Press, Princeton, NJ).

Mantel, Rolf, 1974, On the characterization of aggregate excess demand, Journal of Economic Theory 7.

Mantel, Rolf, 1976, Homothetic preferences and community excess demand functions, Journal of Economic Theory 12, 197–202.

Mas-Colell, Andreu, 1977, Regular, nonconvex economies, Econometrica, 45.

Sonnenschein, Hugo, 1972, Market excess demand functions, Econometrica 40, 549–563.

Sonnenschein, Hugo, 1973, Do Walras identity and continuity characterize the class of community excess demand functions?, Journal of Economic Theory 6, 345–354.

— 4 —

Efficiency and Decentralization in the Pure Theory of Public Goods*

Some basic facts of public goods theory are presented in the primitive set-up of a collection of projects devoid of any linear structure. There is a single private good. Characterizations of Pareto optimal and core states in terms of valuation functions (i.e. supporting "prices") are obtained. Voluntary financing schemes are discussed.

I. Introduction, Model, and Assumptions

I.1. Introduction

It is the intention of this paper to gather and present the basic facts of public goods theory in the primitive setup of a collection of projects devoid of any linear structure (and where, therefore, the notion of "price per unit" is meaningless). The main point to be made is that, in contrast to what could be called the minimal dimension of informational variables, the efficiency and decentralizability results of the theory are quite fundamental and hold with great generality.

There are at least two reasons why we believe that on this occasion the disadvantages of generality do not outweigh the advantages. First, as compared with standard public goods theory, there is no loss of substance. The concept of Lindahl prices, for example, is primarily of theoretical interest,

Quarterly Journal of Economics 94 (1980), 623–641.

* This paper was written in its essential parts during my stay at the Universität Bonn in the academic year 1976–1977. I would like to acknowledge gratefully the financial support of the Sonderforschungsbereich 21.

as it is not devised to model any existing market or even, if one takes the position that markets are in essence mass phenomena, potentially existing ones. Second, it is not uncommon that a public decision problem be given in terms of a choice among a few (say six or seven) projects.

Not much will be found here that is very new or deep. Perhaps the only result that will not sound familiar is part (C) of Proposition 1, which provides a decentralized characterization of the core. Our treatment is connected directly to the work of Dubins [1977] and to the theory of the allocation of public goods, particularly the static theory of Lindahl equilibrium (see Milleron [1972]) and the dynamic mechanisms leading to efficient allocations proposed by Malinvaud [1971] and Drèze-de la Vallée Poussin [1971] (see also Champsaur, Drèze, and Henry [1977]).

Formally our presentation is of the general equilibrium type, but the best interpretation is in partial-equilibrium terms. Indeed, we postulate, as an essential component of the theory, the existence of a single private good that can, and probably should, be thought of as a Hicksian composite commodity (for explicit partial-equilibrium approaches see Groves-Loeb [1975] and Green-Laffont [1977]).

The next subsection describes the model and basic assumptions. The exposition is organized in three parts. Section II presents the analog of "personalized prices" equilibrium theory. Section III considers decision devices obtained by putting "prices" under the control of agents. Section IV iterates the functioning of the decision device.

I.2. The Model and Assumptions

There is given a nonempty, compact, metric space K of *projects* and a finite collectivity of agents $I = \{1, \ldots, n\}$. It is appropriate to think of K as a finite, not very large, set. Similarly, I should also be thought of as not too large. Every agent i has preferences $\succsim_i$ defined on tuples (x,m) of projects and amounts of a unique private good (to be called "money"). The existence of a "money" commodity makes possible, to some extent, the transfer of welfare among agents and it is basic to the approach to social decision theory taken here. It constitutes an interesting specialization of the general theory of social choice (see Mueller [1976] for a recent survey).

A.l. It will *always* be assumed that agents' preferences satisfy: $\succsim_i$ is a continuous, reflexive, complete, transitive preorder on $K \times [0,\infty)$.

A.2. (a) $\succsim_i$ is continuous and strictly monotone in money, i.e., for all $x \in K$ if $m' > m$, then $(x,m') \succ_i (x,m)$.

(b) (Indispensability of money) If $m > 0$, then for any $x, x' \in K$ and $i \in I$, $(x',m) \succ_i (x,0)$.

A.2.(b) is obviously very strong, but it will make our analysis simpler. We let $\{u_i{:}K \times [0,\infty) \to R{:}i \in I\}$ be a family of continuous utility functions for the $\succsim_i$. They will be useful later on. Every $i \in I$ is endowed with a non-negative w_i amount of money.

To complete the description of the model, we introduce a cost function $c{:}K \to (-\infty,\infty]$ that shall be assumed continuous. Note that $c(x) = +\infty$ is allowed; so if $c(x) = +\infty$ and $x_n \to x$, then $c(x_n) \to +\infty$. Occasionally we shall also assume that $c(x) \geq 0$.

Very often we shall postulate the existence of a distinguished project, denoted $0 \in K$, such that $c(0) = 0$. It is to be interpreted as the "status quo," i.e., as the situation from which a change is being contemplated. When there is a $0 \in K$, we assume that $u_i(0,w_i) = 0$ for all $i \in I$.

Definition 1. A *state* is a project $x \in K$ and an assignment of money to agents $m{:}I \to R$. It is denoted (x,m).

Definition 2. A state (x,m) is feasible if

$$c(x) \leq \sum_{i\in I} w_i - \sum_{i\in I} m_i.$$

Definition 3. A state (x,m) is Pareto optimal (P.o.) if it is feasible and if there is no feasible state (x',m') such that $(x',m_i') \succ_i (x,m_i)$ for all $i \in I$.

Under hypothesis A.2 this definition is equivalent to the more usual strong version that allows some agents to remain as well off. The same is true of the next definition.

Definition 4. A state (x,m) belongs to the *core* if it is feasible and there is no $S \subset I$ such that $S \neq \emptyset$ and for some state (x',m'),

$$c(x') \leq \sum_{i\in S} (w_i - m_i')$$

and

$$(x',m_i') \succ_i (x,m_i)$$

for all $i \in S$.

Definition 5. Let there be a status quo $0 \in K$. A state is maximally Pareto improving (m.P.i.) if it is P.o. and $(x,m_i) \succsim_i (0,w_i)$ for all $i \in I$.

Remark 1. The term "maximally Pareto improving" is preferred to the more usual of "individually rational Pareto optimal" because the latter, which is an import from game theory, would be misleading in our context.

II. Valuation Equilibrium

Definition 6. A valuation system is a vector $v = (v_1, \dots, v_n)$ of upper semicontinuous functions v_i: $K \to [-\infty,\infty)$, $i = 1, \dots, n$. We allow $v_i(x) = -\infty$.

From the standpoint of an individual $i \in I$ a valuation function v_i is given as a datum, and it is to be interpreted analogously to prices, i.e., $v_i(x)$ is the amount of money to be relinquished for the right to enjoy the project x. As the source of valuation systems we should think of a coordinating center in charge of announcing and enforcing them. To appeal to impersonal markets, as one does with the usual prices, would not be a sensible thing to do here.

Definition 7. A state $(\bar{x},\bar{m})$ is *supported by a valuation system v* if for some $\Pi = (\Pi_1, \dots, \Pi_n) \in R^n$ with

$$\sum_{i \in I} \Pi_i = \sum_{i \in I} v_i(\bar{x}) - c(\bar{x}):$$

(a) for every i,$(\bar{x},\bar{m}_i)$ maximizes $\succsim_i$ on

$$\{(x, m_i): v_i(x) + m_i = w_i + \Pi_i\},$$

(b) $\bar{x}$ maximizes $\Sigma_{i \in I} v_i(x) - c(x)$ on K.

Parts (a) and (b) are, respectively, utility and profit maximization conditions; Π is a vector of distribution of profits or losses. Note that if $(\bar{x},\bar{m})$ is supported by a valuation system v, then it is supported by a valuation system v' with total profit zero; setting $v'_i(x) = v_i(x) - \Pi_i$ and $\Pi'_i = 0$ will suffice.

Definition 8. A state $(\bar{x},\bar{m})$ is a *valuation equilibrium* if it can be supported by a valuation system.

Note that a valuation equilibrium is automatically a feasible state.

Remark 2. If K is some cube in R^m and a valuation system is restricted to be linear homogeneous, i.e., $v_i(x) = p_i x$, then the concept of a valuation equilibrium coincides with the concept of Lindahl equilibrium (see Milleron [1972] for a survey of the use of this concept in public goods theory). So, valuation systems are a sort of "nonlinear" personalized prices.

Of the following characterization results only (C) can make any claim to novelty.

Proposition 1. (1) The state $(\bar{x},\bar{m})$ is a valuation equilibrium if and only if $(\bar{x},\bar{m})$ is Pareto optimal.

(2) Let there be a status quo project $0 \in K$. The state $(\bar{x},\bar{m})$ is a valuation equilibrium with respect to a valuation system v with $v_i(0) \leq 0$ for all i, if and only if $(\bar{x},\bar{m})$ is maximally Pareto improving.

(3) Let $c(x) \geq 0$ for all $x \in K$. The state $(\bar{x},\bar{m})$ is a valuation equilibrium with respect to a nonnegative valuation system and zero profits, i.e., $\Pi = 0$, if and only if $(\bar{x},\bar{m})$ is in the core.

Proof of Proposition 1. We first verify the only if part of (l)–(3). Take a state (x,m) supported by a valuation system v. Suppose that for some $C \subset I$, $C \neq \emptyset$, there is (x',m') such that $(x',m'_i) \succ_i (\bar{x},m_i)$ for every i and

$$\sum_{i \in C} (w_i - m'_i) \geq c(x').$$

If Π is the vector of distribution of profits, then by (a) and (b) of Definition 7,

$$\sum_{i \in I} v_i(x) - c(x) \geq \sum_{i \in I} v_i(x') - c(x'),$$

$$\sum_{i \in C} (v_i(x') + m'_i) > \sum_{i \in C} (w_i + \Pi_i),$$

and

$$\sum_{i \in C} (w_i - m'_i) - c(x') \geq 0.$$

So, if $C = I$, we have

$$\sum_{i \in I} \Pi_i = \sum_{i \in I} v_i(x) - c(x) > \sum_{i \in I} \Pi_i + \sum_{i \in I} (w_i - m'_i) - c(x') \geq \sum_{i \in I} \Pi_i,$$

a contradiction that takes care of (1). If $\Pi_i = 0$ and $v_i \geq 0$ for all i, then

$$0 = \sum_{i \in I} v_i(x) - c(x) \geq \sum_{i \in I} v_i(x') - c(x')$$

$$\geq \sum_{i \in C} v_i(x') - c(x') > \sum_{i \in C} (w_i - m'_i) - c(x') \geq 0,$$

a contradiction that proves the only if part of (3). The only if part of (2) is an obvious consequence of the above arguments for (1) and the utility maximization hypothesis (b).

We now verify the "if" part. Let $(\bar{x},\bar{m})$ be a feasible state. For every $i \in I$, and $x \in K$, let $g_i(x) = w_i - z$ if $(x,z) \sim_i (\bar{x},\bar{m}_i)$ and $g_i(x) = -\infty$ if no such z exists. By (A.l) if z exists, it is uniquely defined and the function $g_i{:}K \to [-\infty,\infty)$ is continuous. We have $g_i(\bar{x}) = w_i - \bar{m}_i$ and therefore

$$\sum_{i \in I} g_i(\bar{x}) \geq c(\bar{x}).$$

Suppose now that $(\bar{x},\bar{m})$ is P.o. Let us have, for some $x \in K$,

$$a = \sum_{i \in I} g_i(x) - c(x) > 0;$$

then $g_i(x) > -\infty$ for all i. Consider the state (x,m') defined by $m_i' = w_i - g_i(x) + a/n$; since for all i, $(\bar{x},\bar{m}_i) \sim_i (x,w_i - g_i(x))$ we have $(x,m_i') \succ_i (\bar{x},\bar{m}_i)$ for all i, which is impossible because $(\bar{x},\bar{m})$ is P.o. and (x,m') is feasible. We therefore conclude that

$$\sum_{i \in I} g_i(x) \leq c(x)$$

for all $x \in K$. Thus, if we put $v_i = g_i$ and $\Pi_i = 0$, the "if" part of (1) is proved.

Let there be a status quo $0 \in K$ and let $(\bar{x},\bar{m})$ be maximally Pareto improving. Since $(\bar{x},\bar{m}_i) \succsim_i (0,w_i)$ for all $i \in I$, $g_i(0) \leq 0$ for all $i \in I$. Hence, if, as above, we have $v_i = g_i$, we are finished for the "if" part of (2).

Now let $(\bar{x},\bar{m})$ belong to the core. By definition of the core and the functions g_i, we must have

$$\sum_{i \in C} g_i(z) \leq c(x)$$

for all $C \subset I$, $C \neq \emptyset$ and $x \in K$. Then define $v_i{:}K \to [0,\infty)$ by $v_i(x) = \max\{0,g_i(x)\}$ and note that v_i is then nonnegative and continuous. For all $x \in K$,

$$\sum_{i \in I} v_i(x) \leq c(x),$$

since, letting $C = \{i \in I{:}g_i(x) \geq 0\}$,

$$\sum_{i \in I} v_i(x) = \sum_{i \in C} g_i(x) \qquad \text{if } C \neq 0$$

and

$$\sum_{i \in I} v_i(x) = 0 \leq c(x) \qquad \text{if } C = \emptyset.$$

On the other hand,

$$\sum_{i \in I} v_i(\bar{x}) \geq \sum_{i \in I} g_i(\bar{x}) \geq c(\bar{x}).$$

So,

$$\sum_{i \in I} v_i(\bar{x}) = c(\bar{x}),$$

and $\bar{x}$ maximizes profits. Also, for all $i \in I$, if $v_i(x) + m_i \leq w_i$, then $(\bar{x},\bar{m}_i) \succsim_i (x,m_i)$ because $v_i(x) \geq g_i(\bar{x})$ and so, the utility maximization hypothesis is satisfied.

Q.E.D.

Remark 3. Part (3) of Proposition 1 provides a price-like characterization of the core for the case where (A.1)–(A.2) are satisfied and c is non-negative. An analogous characterization for general c could also be obtained. Of course, while maximal Pareto-improving states are easily seen to exist under the assumptions made, the core may well be empty.

III. Voluntary Financing Devices

N.B. In the present section we assume that there exists a status quo point $0 \in K$.

Our point of view will now be changed. Instead of looking at valuation systems as equilibrating parameters controlled by a hypothetical coordinating center, we shall regard them as willingness-to-pay functions under the direct control of the agents.

The voluntary financing public decision method to be described is not new; it was first proposed by Steinhaus [1949] and has recently been extended and discussed by Dubins [1977]. It is built on the old notion of the maximization of social surplus, and it is intimately related to the ideas of Malinvaud [1971] and Drèze-de la Vallée Poussin [1971] concerning dynamic procedures for the allocation of public goods (see Section VI). The name "device" is taken from Dubins.

Definition 9. A *proposal of agent i* is a function $v_i:K \to [-\infty, w_i]$ such that $v_i(0) = 0$.

A proposal is to be interpreted as, for each $x \in K$, the amount $v_i(x)$ that agent *i volunteers* to pay if project *x* is implemented.

Definition 10. A vector $\delta = (\delta_1, \ldots, \delta_i)$ with $\delta_i \geq 0$ and

$$\sum_{i \in I} \delta_i = 1$$

is called a *distribution vector.*

Given a vector of proposals $v = (v_1, \ldots, v_n)$, let K_v be the set of profit-maximizing projects, i.e.,

$$K_v = \left\{ x \in K : \sum_{i \in I} v_i(x) - c(x) \geq \sum_{i \in I} v_i(x') - c(x') \quad \text{for all } x' \in K \right\}.$$

Let Π_v be maximum profits; since

$$\sum_{i \in I} v_i(0) - c(0) = 0,$$

we have $\bar{\Pi}_v \geq 0$.

Definition 11. Given the proposals $v = (v_1, \ldots, v_n)$ and distribution vector $\delta = (\delta_1, \ldots, \delta_n)$, *an outcome* is any state (x,m) where

(i) $x \in K_v$,

(ii) $x \in K_v \setminus \{0\}$ if $K_v \setminus \{0\} \neq \emptyset$,

(iii) $m_i = w_i - v_i(x) + \delta_i \bar{\Pi}_v$.

So, given the stated willingness to pay of the agents, some profit-maximizing state is chosen. Unless there is no alternative, the status quo is not chosen.

N.B. From now on, a fixed distribution vector δ is given and if a triple (v,x,m) is considered, it is to be understood that (x,m) is an outcome for v and δ.

Definition 12. A triple (v,x,m) is *stable* if no $i \in I$ can insure for himself a better outcome by changing his proposal, i.e., if v' is such that $v'_j = v_j$ for $j \neq i$, then for every outcome (x',m'), we have $(x,m_i) \succsim_i (x',m'_i)$.

The word “insure” in the previous definition is justified. Indeed, suppose that for some outcome (x',m'), one had $(x',m'_i) \succ_i (x,m_i)$. By replacing v'_i by a v_i'' with $v_i''(0) = 0$, $v_i''(x') = v_i'(x')$, and $v''(x'') < v'(x'')$ for all $x'' \neq 0,x'$, (x',m') becomes the only possible outcome (i.e., $K_v = \{0\} \cup \{x'\}$).

If the voluntary financing device is viewed as a game, then the notion of stability is akin to the concept of Cournot-Nash equilibrium.

Definition 13. Let (v,x,m) be given. The proposals are called *self-protective* if, for all $x' \notin 0$ and i,

$$(x', w_i - v_i(x') + \delta_i \Pi_v) \succsim_i (x, m_i).$$

The notion of self-protectiveness is akin to the maximin decision rule. For the proposal of agent i to be self-protective with respect to an outcome x, it means that the agent can guarantee himself, independently of the proposals of the other agents, the level of utility corresponding to x and the payment proposed.

Proposition 2. (4) No (v,x,m) with $\bar{\Pi}_v > 0$ can be stable.

(5) If given (v,x,m), v are self-protective and (x,m) is maximally Pareto improving, then (v,x,m) is stable.

(6) Given (v,x,m), if $\bar{\Pi}_v = 0$ and the proposals v constitute a valuation system for (x,m), then (v,x,m) is stable.

Proof of Proposition 2. Claim (4) is quite clear. Take an i with $\delta_i < 1$. Replace v_i by v_i', where $v_i'(0) = 0$, $v_i'(x) = v_i(x) - \bar{\Pi}_v$, $v_i'(x') < v_i(x') - \bar{\Pi}_v$ for $x' \neq 0,x$. The function v' can be chosen to be continuous if so desired. Then (x, m') is an outcome for v' such that $\bar{\Pi}_{v'} = 0$, $m_j' = m_j$ for $j \neq i$ and

$$m_i' = w_i - v_i'(x) = w_i + \bar{\Pi}_v - v_i(x) > w_i + \delta_i \Pi_v - v_i(x) = m_i,$$

i.e., i is better off.

For claim (5), suppose that it was not true. Then there would be an agent i who could insure himself a better outcome. But since proposals are self-protective, none of the other agents can be worse off with the new outcome. By hypothesis (A.2), the outcome could not be maximally Pareto improving.

To prove (6), consider any agent i. The outcomes that i can possibly enforce are of the form $(x',w_i - m_i')$, where

$$m_i' \geq c(x') - \sum_{j \neq i} v_j(x')$$

But

$$\sum_{j \in I} v_j(x') \leq c(x')$$

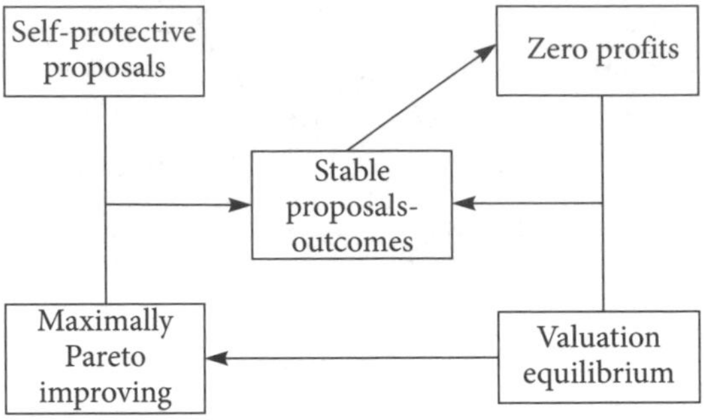

Figure I

and so $m_i' \geq v_i(x')$. Since v is a valuation equilibrium for (x,m), $(x,w_i - v_i(x)) \succsim_i (x',w_i - v_i(x')) \succsim_i (x',w_i - m_i')$, i.e., agent i cannot change his proposal to advantage.

Q.E.D.

Observe that if, with respect to (x,m), proposals v are self-protective and have the valuation property, then for all i and $x' \neq 0$, either $v_i(x') = w_i - z$, where $(x',z) \sim_i (x,m_i)$, or $v_i(x') = -\infty$ if no such z exists.

Figure I illustrates the interrelationships among the different concepts.

It is only for simplicity that in the definition of stability (Definition 12) we allowed departures only by individuals. Every result remains valid if, in addition we had required stability against departures by whole groups of individuals.

Now suppose that a voluntary financing device is put into operation in order to decide among a set of projects.

It is natural to surmise that prior to the emergence of the definitive proposals, there will be a more or less structured "negotiation period" where proposals are tentatively put forward by agents and then possibly revised. The analysis of negotiation behavior falls into the domain of game theory, and clear-cut deterministic rules cannot be expected from it. The position we shall take here is that, through some mixture of conflict and cooperation, bargaining will eventually lead to proposals yielding maximally Pareto-improving outcomes. In other words, *for the interpretation at hand* (not many projects and not many individuals) we accept the principle that if there are, for all, obvious and easily reachable gains from cooperation, bargaining will not get definitely stuck at proposals with very inefficient outcomes (someone will, sooner or later, break a deadlock by volunteering a larger contribution).

A different matter is the stability of the final agreement. If the final proposals are not stable, i.e., mutually reinforcing, there is an incentive on the part of some agents to break the agreement, i.e., "free riders" will appear and the agreed-on outcome will decompose or perhaps, simply, it will never quite be reached. The gains from cooperation cannot be consolidated. The plausibility and resoluteness of cooperative bargaining will therefore be particularly good if it leads to proposals, which besides giving maximally Pareto-improving outcomes, do not introduce enforcement problems, i.e., which are stable.

In the next section we present a very natural cooperative iteration procedure of the voluntary financing device that is utility monotone and leads to proposals and outcomes that are self-protective and have the valuation property. So, the proposals are stable, and the outcomes are maximally Pareto efficient. Thus, the voluntary financing device has the potential for quite a good performance by the standards of the two previous paragraphs. The procedure is the analog in our context of the mechanisms of Malinvaud and Drèze-de la Vallée Poussin.

There is another, more demanding, approach to the incentive problem that is associated with the names of Hurwicz [1972] and Groves-Ledyard [1977]. They formalize allocation procedures as *games* with messages as strategies. Further, the games are viewed as *noncooperatiue* and as a solution the notion of *Nash equilibrium* is appealed to. A procedure is good, incentive-wise, if *all* the Nash equilibria of the noncooperative game are Pareto optimal ("all" because there are no grounds to discriminate among them). While this may be an appropriate standard of adequacy in particular cases, one may wonder whether its imposition as a general rule is not too rigid a requirement. In the first place it can be questioned whether incentive problems are always best modeled as games, with its implied sophisticated and foresighted behavior on the part of agents. In the second place, one would feel that bargaining has as many elements of cooperation as of pure conflict so that a strictly noncooperative individualistic setup is quite limited. Rigorously speaking, a game is noncooperative if there are no preplay communication possibilities whatsoever. This is a very specific restriction for so general a problem as the one we are discussing. In the third place, the presumption that the process of noncooperative bargaining will come to rest at a Nash equilibrium is justified only in situations with a large number of agents; otherwise the Cournot-Nash conjecture is unrealistic.

At any rate, we should mention that the voluntary financing device does not have the potential to meet the desiderata of the last paragraph, as it is quite clear there may be plenty of stable proposals with outcomes that are very far from being Pareto optimal.

IV. A Voluntary Financing Process

The problem of this section is how, by means of a voluntary financing device, can we reach triples (v,x,m) where v are stable proposals and (x,m) is maximally Pareto efficient. We adopt the most extreme cooperative outlook and except for the stability requirement on v, incentive problems are put aside. See the end of the previous section for a discussion of those.

If individual utility functions were separable and linear in money, i.e., of the form $u_i(x,m_i) = \hat{u}_i(x) + k_i m_i$, then an obvious device is the maximization of surplus, i.e., we put $v_i = \hat{u}_i$. The result is a maximally Pareto-optimal outcome and self-protective, hence stable, proposals. Thus, we have a one-shot functioning device. This suggests that for the general case (i.e., when "income effects" exist), we try an iteration process where each step amounts to a maximization of surplus.

IV.1. Definition and Convergence Properties of the Process

N.B. In this subsection the distribution vector δ is given.

Without loss of generality we can assume that the range of the utility functions u_i (see Section I) is $[0,\infty)$.

For every i and $r \geq 0$ define $g_i^r : K \to [w_i, -\infty]$ by the solution to $r = u_i(x, w_i - g_i^r(x))$ if it exists. Otherwise $g_i^r(x) = -\infty$.

Define an iterative process $(v_0,x_0,m_0), \ldots, (v_t,x_t,m_t), \ldots$ as follows:

1. Take $x_0 = 0$, $v_{i0}(x) = 0$ all $x \in K$, $m_{i0} = w_i$.

2. Let (v_t,x_t,m_t) be given and suppose that $u_t \geq u_{t-1} \geq \cdots \geq 0$, where $u_{it} = u_i(x_t,m_{it})$. Then put $v_{i,t+1}(x) = g_i^{u_{it}}(x)$ if $x \neq 0$ and $v_{i,t+1}(0) = 0$. If $K_{v_{t+1}} = \{0\}$, take $x_{t+1} = 0$; if $K_{v_{t+1}} \backslash \{0\} \neq \emptyset$, take for x_{t+1} any project in the set $K_{v_{t+1}} \backslash \{0\}$. Of course, we let $m_{i,t+1} = w_i + \delta_i \bar{\Pi}_{v_{t+1}} - v_{i,t+1}(x_{t+1})$. Every $v_{i,t+1}$ is upper semicontinuous, hence $K_{v_{t+1}} \neq \emptyset$, and the process is well defined. By construction the proposals v_{t+1} are self-protective for (x_t,m_t) so that, indeed, $u_{t+1} \geq u_t$.

Let J be the (nonempty) set of limit points of $\{x_t\}$ and Π_t total profits at stage t.

Proposition 3. (7) $\Pi_t \to 0$, the sequence (v_t,m_t) converges to a limit $(\bar{v},\bar{m})$ and for any $\bar{x} \in J$, $\bar{v}$ is a valuation system supporting $(\bar{x},\bar{m})$. Hence $(\bar{x},\bar{m})$ is maximally Pareto improving, and $(\bar{v},\bar{x},\bar{m})$ is stable.

Proof of Proposition 3. Note that the sequence u_t is monotone increasing and bounded above. Hence $u_t \to \bar{u}$ for some $\bar{u} \in R^n$. This, the compactness of K, and the strict monotonicity of preferences with respect to money imply that $\Pi_t \to 0$. By the continuity of preferences and the definition of v_t we must have $v_t \to \bar{v}$ for some proposals $\bar{v}$ and, therefore, $m_t \to \bar{m}$ for some $\bar{m} \in R^n$. For any t and i, (x_t,m_{it}) maximizes $\succsim$ on

$$\{(x',m'):v_{i,t+1}(x') + m' \leq w_i + \delta_i\Pi_t\}.$$

The profit maximization condition is also satisfied at every t. Hence, by continuity, $\bar{v}$ supports $(\bar{x},\bar{m})$ for any $\bar{x} \in J$.

Q.E.D.

Remark 4. If K is infinite, J may well not be unique. If K is finite, then there is a finite number of steps after which $u_t = \bar{u}$ and, therefore, $v_t = \bar{v}$ for all t. Indeed, suppose that $u_{t+1} > u_t$ infinitely often, then $x_{t+h} = x_t$, and $u_{t+h} > u_t$ for some t and h, and so $m_{t+h} > m_t$. Hence,

$$c(x_{t+h}) - \sum_{i \in I} w_i = \sum_{i \in I} m_{i,t+h} > \sum_{i \in I} m_{i,t} = c(x_t) - \sum_{i \in I} w_i,$$

which is a consideration. Summing up: the process can be stopped after a finite number of iterations.

Remark 5. The process just presented is nothing but a global version of the procedures proposed by Malinvaud [1971] and Drèze-de la Vallée Poussin [1971] for the allocation of public goods. Naturally, the convergence and incentive properties are also the same. Let $K \subset R^m$ be some large convex cube and suppose that the utility functions are smooth. Then at step t instead of going to the global maximum of the profit function (which, incidentally, would require whole functions as proposals) we could simply follow the direction of the gradient. The process obtained thus would belong to the Malinvaud-Drèze-de la Vallée Poussin family, and, overall, would be much more economical in information using and processing.

IV.2. Neutrality (unbiasedness) Properties of the Process

Neutrality is Champsaur's term [1976], unbiasedness, Hurwicz's [1959]. Both are coined to denote the mechanism that, possibly depending on the

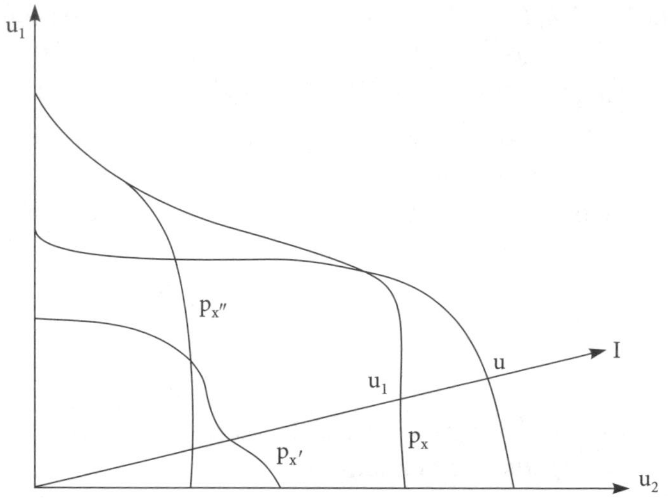

Figure II

specification of some parameter, can reach every Pareto-optimal point (or perhaps, every maximally Pareto-improving one).

Let $P \subset R^n_+$ be the Pareto set in nonnegative utility space. Remember that 0 is the utility of the status quo point.

For every project $x \in K$, let $P_x \subset P$ be the utility vectors in P that are feasible with the constraint that project x is in fact implemented. So, if x is at all feasible,

$$P_x = \left\{ (u_1(x, m_1), \ldots, u_n(x, m_n)) \in P : \sum_{i \in I} m_i = \sum_{i \in I} w_i - c(x) \geq 0 \right\},$$

and P is the upper frontier of $\cup_{x \in K} P_x$.Because of (A.2) every P_x and, therefore, also P has the property that it intersects any nonnegative ray through the origin of R^n at exactly one point. So, P_x,P are, up to homeomorphism, simplices. See Figure II.

Consider now the voluntary financing process of the previous subsection. Given the distribution vector δ, let $P(\delta)$ be the utility vectors to which the process can lead. Given the possible indeterminacy if there is more than one profit-maximizing project, $P(\delta)$ need not be a singleton. Let $V = \cup_{\delta \in \Delta} P(\delta)$, where Δ is the $n - 1$ unit simplex.

For the Malinvaud-Drèze-de la Vallée Poussin process, Champsaur [1976] showed that $V = P$. This is the property of neutrality, or unbi-

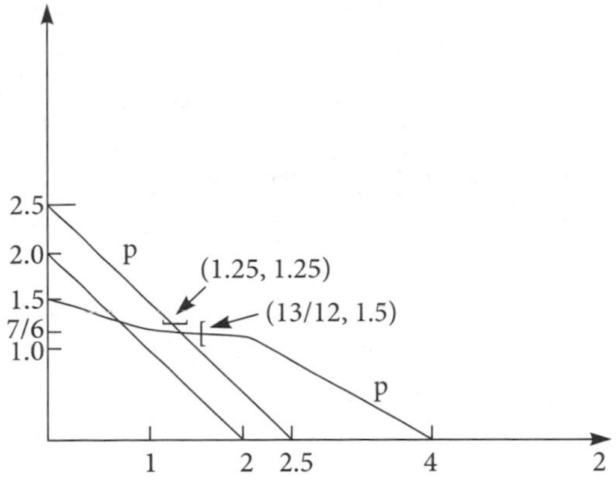

Figure III

asedness. Given the nonconvexities of our problem, it should not be surprising that

(8) $V = P$ may not hold.

Example. $K = \{0, b, c, d\}$, and there are two agents with $w_1 = w_2 = 1$. The cost is zero for every project. The utility functions are

$$u_1 = \begin{cases} u_1(0,m) = 1 - m \\ u_1(b,m) = m \\ u_1(c,m) = 2m - 1 \\ u_1(d,m) = \begin{cases} 2m - \dfrac{1}{2} & m \leq \dfrac{3}{4} \\ \dfrac{2}{3}m + \dfrac{1}{2} & m \geq \dfrac{3}{4} \end{cases} \end{cases}$$

$$u_2 = \begin{cases} u_2(0,m) = 1 - m \\ u_2(b,m) = m \\ u_2(c,m) = 2m - \dfrac{1}{2} \\ u_2(d,m) = \begin{cases} 2m - 1 & m \leq 1 \\ 4m - 3 & m \geq 1. \end{cases} \end{cases}$$

The Pareto set is graphed in Figure III. By a little tedious computing, it can be verified that when $\delta_1 > \delta_2$, agent 1 gets utility larger than 1.25, while if $\delta_2 > \delta_1$, agent 2 gets more than 1.5. At $\delta_1 = \delta_2$ two final outcomes are possible: $(u_1,u_2) = (1.25,1.25)$, or $(u_1,u_2) = (13/12,1.5)$. The "open segment" in $P\,((1.25,1.25), (13/12,1.5))$ is therefore never reached.

Remark 6. It is not hard to verify that an example of nonneutrality must involve at least three projects, not counting the status quo.

Suppose that we drop from the specification of the voluntary financing procedure the requirement that at every step profits are redistributed according to a fixed predetermined vector δ, and let us leave this distribution indeterminate, i.e., any redistribution is admissible. Denote then by V' the utility vectors to which the process can lead. Trivially, we have

Proposition 4.

(9) $V' = P$.

Proof of Proposition 4. This is very clear. Pick any $u \in P$ and let I be a half ray through u. Since the process converges to Pareto optimally and the iteration rules are the same at every step, it suffices to show that at step 1 profits can be redistributed so that $u_1 \in I$. Let x_1 be profit-maximizing at step 1. Then Px_1 intersects I between 0 and u. Let u_1 be the intersection point (see Figure II). Distribute profits at step 1 so as to reach u_1.

Q.E.D.

University of California, Berkeley

References

Champsaur, P., "Neutrality of Planning Procedures in an Economy with Public Goods," *Review of Economic Studies,* XLIII (1976), 293–99.

———, J. Drèze, and C. Henry, "Stability Theorems with Economic Applications," *Econometrica,* XLV (1977), 273–94.

Drèze, J., and D. de la Vallée Poussin, "A Tâtonnement Process for Public Goods," *Review of Economic Studies,* XXXVIII (1971), 133–50.

Dubins, L., "Group Decision Devices," *American Mathematical Monthly,* LXXXIV (1977), 350–56.

Green, J., and J.-J. Laffont, "Characterization of Satisfactory Mechanisms for the Revelation of Preferences for Public Goods, *Econometrica,* XLV (1977), 472–538.

Groves, T., and M. Loeb, "Incentives and Public Inputs," *Journal of Public Economics,* IV (1975), 211–26.

Groves, T., and J. Ledyard, "Optimal Allocations of Public Goods: A Solution to the Free Rider Problem," *Econometrica,* XLV (1977), 783–810.

Hurwicz, L., "Optimality and Informational Efficiency in Resource Allocation Processes," in K. Arrow, S. Karlin, and P. Suppes, eds., *Mathematical Methods in the Social Sciences* (Stanford, CA: Stanford University Press, 1959).

———, "On Informationally Decentralized Systems," in R. Radner and B. McGuire, eds., *Decision and Organization* (Amsterdam: North-Holland, 1972).

Malinvaud, E., "A Planning Approach to the Public Goods Problem," *Swedish Journal of Economics,* I (1971), 96–111.

Milleron, J. C., "Theory of Value with Public Goods: A Survey Article," *Journal of Economic Theory,* V (1972), 419–77.

Mueller, D., "Public Choice: A Survey," *Journal of Economic Literature,* XIV (June 1976).

Steinhaus, H., "Sur la division pragmatique," *Econometrica,* supplement, XVII (1949), 315–19.

— 5 —

The Price Equilibrium Existence Problem in Topological Vector Lattices[1]

A price equilibrium existence theorem is proved for exchange economies whose consumption sets are the positive orthant of arbitrary topological vector lattices. The motivation comes from economic applications showing the need to bring within the scope of equilibrium theory commodity spaces whose positive orthant has empty interior, a typical situation in infinite dimensional linear spaces.

Keywords: Walrasian equilibrium, price equilibrium, existence, topological vector lattice, Riesz space, proper preferences.

1. Introduction

In this paper we reconsider the price equilibrium existence problem for exchange economies with an infinite number of commodities and finitely many consumers. Our purpose is to do so in a context sufficiently general to encompass as particular instances a number of commodity spaces that have been found useful in applications. For example: (i) $L_\infty(M, \mathscr{M}, \mu)$, or L_∞ for short, the space of essentially bounded measurable functions on a measure space (see Bewley (1972)), which is relevant to the analysis of

Econometrica 54 (1986), 1039–1053.

1. This paper was first circulated in February of 1983 under the title: "The Prices of Equilibrium Existence Problem in Banach Lattices." It has been presented to a number of seminars and conferences and I am most indebted to their audiences. Thanks are due to C. Aliprantis, who has been most patient in answering my mathematical queries, C. Huang, C. Ionescu-Tulcea, L. Jones, W. Zame, and two anonymous referees. Financial support from the National Science Foundation is gratefully acknowledged.

allocation of resources over time or states of nature; (ii) $ca(K)$, the space of countable additive signed measures on a compact metric space (see Mas-Colell (1975), and Jones (1984a)), which has been exploited for the analysis of commodity differentiation; (iii) $L_2(M, \mathcal{M}, \mu)$, the space of square integrable functions on a measure space (cf. Duffie-Huang (1983a), Chamberlin-Rothschild (1983), Harrison-Kreps (1979)), which arises in finance economics.

For a survey of the finite dimensional existence theory, see Debreu (1982). The classical reference for the infinite dimensional theory is Bewley (1972). Mathematically, a crucial feature of his analysis (and of his generalizations (Bojan (1974), El'Barkuki (1977), Toussaint (1984), Florenzano (1983), Ali Khan (1984), Yannelis-Prabhakar (1983)) is that the consumption set, which is the natural positive orthant of the commodity space, L_∞ in his case, has a nonempty norm-interior. It is on account of this that spaces such as $ca(K)$ or L_2 are not covered by his work and that a more general, unifying approach is called for. Jones (1984a) has offered an existence theorem for $ca(K)$. We mention, as an incidental remark, that the initial stimulus for our research was an attempt to understand the role of the differentiability-like hypotheses that in contrast to Bewley (1972) appear in Jones (1984a): are they required to guarantee the existence of a price functional or to yield one with a particular important property, i.e., continuity on characteristics? Another thought-provoking paper was Ostroy (1984) where L_1 spaces are considered.

The mathematical setting for our commodity spaces will be *topological vector lattices.* Those turn out to be sufficiently general for our aims and also tractable. Commodity spaces which are vector lattices, or Riesz spaces, have been introduced by Aliprantis and Brown (1983) in the context of an excess-demand approach to equilibrium. A topological vector lattice is a vector lattice where the lattice operations $(x, y) \mapsto x \vee y$ are (uniformly) continuous. A most important example are the *Banach lattices.* Those are Banach spaces equipped with a positive orthant which induces a lattice order $\geqslant$ with the property that $|x| \leqslant |y|$ implies $||x|| \leqslant ||y||$. Here $|x| = (x \vee 0) - (x \wedge 0)$ is the absolute value of x. See Aliprantis and Burkinshaw (1978), and Schaefer (1971). All the examples mentioned above (and many more, e.g., all the l_p and L_p spaces) are Banach lattices when equipped with their natural norms and orders. A Banach space which is not a Banach lattice is $C^1([0, 1])$. A vector lattice which is not a Banach space is R^∞. A major surprise of this paper is precisely this relevance of lattice theoretic properties to the existence of equilibrium problem. One would not have been led to expect it from the finite dimensional theory. In the latter it is possible to formalize and solve the existence problem using only the topological and convexity structures of the space (cf. Debreu (1962)). Informally, a source for the difference seems to be the following: in general vector lattices, order intervals (i.e., sets of the form $\{x\colon a \leqslant x \leqslant b\}$) are, as convex sets, much more tractable and well behaved than general bounded, convex sets. An unfortunate consequence of this is that the exchange case, the one we handle, is inherently simpler than the general production case (see Section 9).

As for the technique of proof, we follow the Negishi (1960) approach and carry our fixed point argument on the utility possibility frontier. This is particularly sensible in a model with infinitely many commodities but only a finite number of consumers. For L_∞ this line of attack was proposed and implemented by Bewley in a regretably unpublished paper (1969), and then by Magill (1981). Their papers have been quite helpful to our analysis. Notably, Bewley already asserted that "it can be expected that this method of proof may be applied to infinite dimensional commodity spaces other than L_∞."

The paper, and the proof of the eventual result, is divided in four parts which we now describe briefly. Each part tackles a conceptually distinct issue. It is worth noting that the first three are trivial in the finite dimensional case.

After short sections on commodities and prices, Section 4 takes care of a preliminary technical and conceptual step, namely, it attacks and solves, rather straightforwardly, the one consumer problem. Only separation arguments are involved.

Sections 5 and 6, which involve no essentially new idea, define and study the properties of the utility possibility set of the economy; in particular, its closedness.

Section 7 is the heart of the paper. We give conditions on individual preferences and the commodity space for weak optima to be supportable by nontrivial price systems. It should be emphasized that this does not follow automatically from the well-behavedness of single consumer problems (i.e., from the solution in Section 4). It is in this section, and only in this section, that the lattice properties of the space are exploited.

Finally, Section 8 states a quasiequilibrium existence theorem and carries out the fixed point proof in the utility possibility frontier. The arguments are identical to those of Bewley (1969) and Magill (1981).

Section 9 mentions possible extensions and comments on some of the recent literature.

2. The Commodity Space

The commodity space, denoted L, is a (real) vector space endowed with two structures: a linear order, denoted $\geqslant$ and a linear topology. We impose the weak regularity conditions that the topology be Hausdorff and locally convex and that the positive cone of the space $L_+ = \{x \in L: x \geqslant 0\}$ be closed. Technically: L is a locally convex, Hausdorff *ordered vector space.*

This formalization is general enough to cover every case of interest. More restricted cases of commodity spaces will be considered in subsequent sections.

3. Price Functionals

We shall take as price valuation functions the elements of L^*, i.e., the continuous linear functionals on L, This is very general. It is worth noting that economic consideration will often lead us to positive linear functionals and that in some spaces of interest (e.g., Banach lattices, see Section 7) those are automatically continuous (see Schaefer (1974, p. 84), for Banach lattices).

4. The Single Consumer Economy

Suppose we have a single consumer characterized by the *consumption set* $L_+ = \{x \in L: x \geqslant 0\}$, the *endowment vector* $\omega \in L_+$ and a *preference relation* $\succsim$ on L_+ which we take to be *reflexive, transitive, complete, continuous* (i.e., the set $\{(x, y) \in L_+ \times L_+ : x \succsim y\}$ is closed), *convex* (i.e., the set $\{y \in L: y \succsim x\}$ is convex for every $x \in L$), and *monotone* (i.e., for every $x \in L_+, y \geqslant x$ implies $y \succsim x$ and there is $v \geqslant 0$ such that $x + \alpha v \succ x$ for all $\alpha > 0$).

With the above hypothesis it is well known that in the finite dimensional case, i.e., when $L = R^n$, the vector ω can always be sustained as a *price quasiequilibrium*, i.e., there is $p \neq 0$ such that $z \succsim \omega$ implies $p \cdot z \geqslant p \cdot \omega$. The next example shows that this ceases to be true in the infinite dimensional case.

Example 1: Let $K = Z_+ \cup \{\infty\}$ be the compactification of the positive integers. Our commodity space will be $L = ca(K)$, i.e., the space of signed bounded countably additive measures with the bounded variation norm. For $x \in L$ and $i \in K$ let $x_i = x(\{i\})$.

For every $i \in K$ define a function $u_i : [0, \infty) \to [0, \infty)$ by

$$u_i(t) = \begin{cases} 2^i t & \text{for } t \leqslant \dfrac{1}{2^{2i}}, \\ \dfrac{1}{2^i} - \dfrac{1}{2^{2i}} + t & \text{for } t > \dfrac{1}{2^{2i}}. \end{cases}$$

See Figure 1.

The preference relation $\succsim$ on the consumption set L_+ is then given to us by the concave utility function $U(x) = \sum_{i=1}^{i=\infty} u_i(x_i)$. It is an easy matter to verify that $U(x)$ is well defined, strictly monotone (i.e., $x \geqslant y$, $x \neq y$ implies $x \succ y$) and weak-start continuous (i.e., U is continuous for the weak convergence for measures).

Let now $\omega \in L_+$ be defined by $\omega_i = 1/2^{2i+1}$ for $i < \infty$ and $\omega_\infty = 1$. We claim that the only linear functional p such that $p \cdot z \geqslant p \cdot \omega$ whenever $z \succsim \omega$ is the trivial $p = 0$. This is a consequence of the unboundedness of the sequence of marginal utilities $u_i'(\omega_i) = 2^i$, $i < \infty$. Indeed, let p be such a functional. For any $x \geqslant 0$ we have $\omega + x \geq \omega$. Hence, $p \cdot x \geqslant 0$, i.e., p is a positive linear functional. For $i \in K$ denote $p_i = p \cdot e_i$, where $e_i(\{j\}) = 1$ if $j = i$ and $= 0$ otherwise. Suppose first that $p \cdot \omega > 0$. Then, by the usual marginal rate of substitution arguments, we should have $p_1 > 0$ and $p_i =$

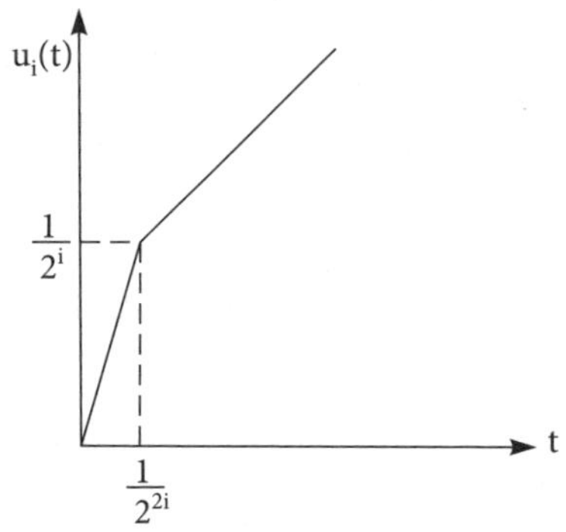

Figure 1

$p_1 z^{i-1}$, $i < \infty$, $p_\infty = \frac{1}{2} p_1$. Define $z \in L_+$ by $z_i = 1/p_i$ and $z^n \in L_+$ by $z_i^n = z_i$, if $i \leqslant n$, and $z_i^n = 0$ otherwise. Then $z - z^n \geqslant 0$ which implies $p \cdot z^n \leqslant p \cdot z$ for all n. But $p \cdot z^n = \Sigma_{i=1}^n p_i z_i = n > p \cdot z$ for n sufficiently large. Contradiction. The only possibility left is $p \cdot \omega = 0$. Pick an arbitrary $z \geqslant 0$ and let $z^m \in L_+$ be defined by $z_i^m = z_i$ if $m < i < \infty$ and $z_i^m = 0$ otherwise. For every m we have $0 \leqslant \alpha(z - z^m) \leqslant \omega$ for some $\alpha > 0$. Hence $p \cdot (z - z^m) = 0$ for all m. The sequence z^m converges to 0 in the (variation) norm on $ca(K)$. For this linear topology in $ca(K)$ any positive linear functional is continuous (see Section 3). Therefore, $p \cdot z^m \to 0$ and so $0 = \lim_m p \cdot z^m = p \cdot z$. Because $L = L_+ - L_+$ we conclude $p \cdot z = 0$ for any $z \in L$, i.e., $p = 0$.

Obviously, if positive results are to be obtained for N consumer economies, they will have to apply to the one consumer case. So the situation of the previous example has to be ruled out. *We shall do so, essentially, by assumption.* Specifically, and informally, we shall impose a priori bounds on the marginal rates of substitution displayed by admissible preference relations with respect to a given composite bundle of commodities.

The next definition can be looked at as a strengthening of monotonicity.

Definition: The preference relation $\succsim$ is *proper* at $x \in L_+$ if there is $v \geqslant 0$ and an open neighborhood of the origin V with the property that $z \in L$ and

$$x - \alpha v + z \succsim x \text{ and } \alpha > 0 \text{ implies } z \notin \alpha V.^*$$

We say that $\succsim$ is *uniformly proper* if $\succsim$ is proper at every $x \in L_+$ and v, V can be chosen independently of x.

As indicated, properness at x merely says that the improvement direction v is desirable in the sense that it is not possible to compensate for a loss of αv with a vector which is too small relative to αv. Observe that the vectors z such that $x - \alpha v + z \notin L_+$ do not create any problem. They pass the test vacuously.

Geometrically, properness at x means that there is an open cone $\Gamma \subset L$ containing a positive vector (hence Γ is nonempty) such that $(-\Gamma) \cap \{z - x \in L_+ : z \succsim x\} = \varnothing$. See Figure 2. Properness at x is automatically satisfied if there is a continuous, positive functional p such that $p \cdot z \geqslant p \cdot x$ whenever $z \succsim x$. Indeed, take $v \geqslant 0$ and V such that $p \cdot v > 0$ and $z \in V$

* Corrects the journal publication.

implies $|p \cdot z| < |p \cdot v|$. Conversely, if $\succsim$ is convex, then properness at x implies the existence of such a functional (see Section 7). Thus, as intended, in the convex case properness at a point is equivalent to the property we want to have. Uniform properness requires that the same cone work at all x.

If L_+ has a nonempty interior (e.g., $L = L_\infty$ with the sup. norm), then uniform properness is automatically implied by monotonicity. It may also happen that, at least at a point, the property is implied by monotonicity, convexity, and continuity; it follows from Bewley (1972) that this is the case for strictly positive points of L_∞ with the Mackey topology. But in general Example 1 has established that properness at a point is an additional hypothesis. The relation among properness, uniform properness and the existence of supporting prices is expanded in Section 7.

The properness property has precedents. Debreu and Hildenbrand (see Bewley (1972)) suggested the use of a boundedness condition on marginal utilities. In the context of an optimal growth problem in Hilbert space, Chichilnisky-Kalman (1980) imposed a condition analogous to properness at a point. Of course, Jones (1984a) should also be mentioned. Simultaneously and independently of us, a similar tool has been used by Ostroy (1984). It appears that our property is intimately related to the extendability of the preference relation $\succsim$ to the entire space as a continuous, convex, monotone preference relation. See Richards (1985a, 1985b) for this. Extendible preferences have been used in Chichilniski- Heal (1984) and Mas-Colell (1985b).

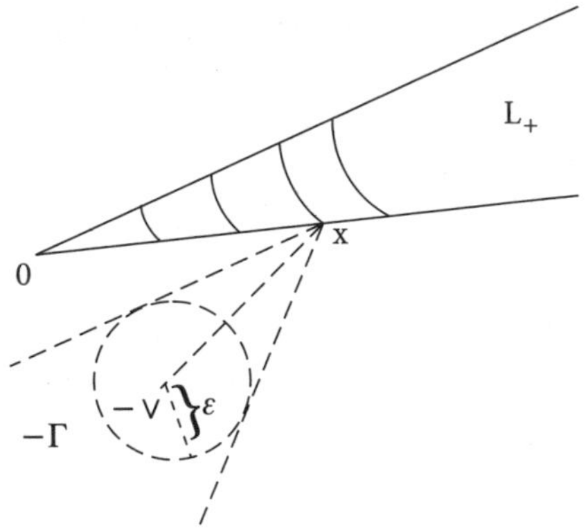

Figure 2

5. A Digression on Utility Theory

As a topological space, L_+ need not be second countable. Our two reference spaces L_∞ and $ca(K)$ are not so in general. Thus even if the preference relation $\succsim$ on L_+ is continuous, we cannot appeal to Debreu's theorem (1954) in order to represent $\succsim$ by a (continuous) utility function. However, the monotonicity of $\succsim$ can be exploited for that purpose in the manner of Kannai (1970). Because continuous utility representations are extremely convenient technical tools, we digress to present such a result.

Proposition 1: *Let $\succsim$ be a continuous preference relation on a subset X of an ordered vector space L of the form $X = \{x \in L: a \leqslant x \leqslant b\}$ for two fixed $a, b \in L$. Suppose that for any $x, y \in X$, $x \geqslant y$ implies $x \succsim y$. Then there is a continuous utility function $u : X \to [0, 1]$ such that $x \succsim y$ if and only if $u(x) \geqslant u(y)$.*

Proof: This follows by standard arguments in utility theory. If $X = \emptyset$ or $a \succsim b$ there is nothing to prove. So, let $b \geqslant a$ and $b \succ a$. Consider first the set $J = \{\alpha a + (1 - \alpha)b: 0 \leqslant \alpha \leqslant 1\}$. Because J is topologically a segment there is a continuous $f : J \to [0, 1]$ such that $f(0) = 0, f(1) = 1$ and f represents the restriction of $\succsim$ to J. For any $x \in X$ let $v(x)$ be such that $v(x) \in J$ and $v(x) \sim x$. Because J is connected and a subset of the union of closed sets $\{z: z \succsim x\}$, $\{z: x \succsim z\}$, such a $v(x)$ must exist. Let then $u(x) = f(v(x))$. Obviously u is a utility function. To see that it is continuous take any $t \in [0, 1]$. Since $u(X) = [0, 1]$ there is $x \in X$ such that $u(x) = t$. Then the sets $u^{-1}([t, \infty)) = \{z: z \succsim x\}$, and $u^{-1}((-\infty, t]) = \{z: x \succsim z\}$ are closed by the continuity of $\succsim$. *Q.E.D.*

The above Proposition is all we shall need. It does not imply, however, that a $\succsim$ defined on L_+ admits a utility function on the entire L_+. See Monteiro (1985) and Shafer (1984) for recent investigations on utility theory for continuous, monotone preferences in linear ordered spaces.

6. Efficient Allocations and the Pareto Frontier

Suppose we have N consumers with preferences $\succsim_i$ on the consumption set L_+. There is also a total endowment vector ω. We assume that every $\succsim_i$ is continuous and monotone. To avoid degeneracy, we take ω to be desirable, i.e., $\alpha\omega \succ_i 0$ for all $\alpha > 0$ and $i \in N$.

Let $X = \{z \in L_+ : 0 \leqslant z \leqslant \omega\}$. A vector $x \in X^N$ such that $\sum_{i=1}^{N} x_i \leqslant \omega$ is called a *feasible allocation*. Denote by $\hat{X} \subset X^N$ the set of feasible allocations.

By Proposition 1 there is, for every $i \in N$, a continuous utility function $u_i: X \to [0, 1]$ with $u_i(0) = 0$, $u_i(\omega) = 1$. Define $U: X^N \to [0, 1]^N$ by $U(x) = (u_1(x_1), \ldots, u_N(x_N))$. Denote by Δ the closed $N - 1$ simplex and, for any $s \in \Delta$, let $f(s) = \sup \{\alpha \in R: \alpha s \in U(\hat{X})\}$. Then:

Proposition 2: $f: \Delta \to [0, N]$ is a well-defined continuous function. Moreover, $f(s) > 0$ for all $s \in \Delta$.

Proof: $f(s) > 0$ for all $s \in \Delta$ is an obvious consequence of the desirability of ω.

Let $s_n \to s$ and $\alpha s = u(x)$ for some $x \in \hat{X}$ and $\alpha > 0$. Pick $0 < \beta < \alpha$ and suppose, without loss of generality, that $\beta s_{ni} < \alpha s_i$ for all n and any $i \in N$ with $s_i > 0$. For any i with $s_i = 0$ let $u_i(\delta_{ni}\omega) = \beta s_{ni}$. Put $\delta_n = \Sigma_{s_i=0}\, \delta_{ni}$. Then $\delta_n \to 0$. Define an allocation y_n by $y_{ni} = \delta_{ni}\omega$ if $s_i = 0$ and $y_{ni} = (1 - \delta_n)x_i$ otherwise. We have that, for sufficiently large n, $u_i(y_{ni}) \geqslant \beta s_{ni}$ for every i. So we have found $y \in \hat{X}$ such that $U(y) \geqslant \beta s_n$ componentwise. Let now $\mu_i \in [0, 1]$ be such that $u_i(\mu_i y_i) = \beta s_{ni}$. Then $\Sigma_{i \in N} \mu_i y_i \leqslant \Sigma_{i \in N} y_i \leqslant \omega$; i.e., we have a feasible allocation which has a utility image that is precisely βs_n. Since $\beta < \alpha$ is arbitrary we conclude that $\lim_{s_n \to s} \inf f(s_n) \geqslant f(s)$.

The proof that $\lim_{s_n \to s} \sup f(s_n) \leqslant f(s)$ is analogous and even simpler.

Q.E.D.

With no change in the proof, the above proposition remains valid if preferences are restricted to consumption sets which are closed, convex subsets of L_+ containing the origin and ω.

A feasible allocation $x \in X^N$ is *weakly efficient* if there is no other feasible allocation x' such that $x'_i \succ_i x_i$ for every $i \in N$. If x is weakly efficient then $U(x)$ must belong to the upper frontier of $U(\hat{X})$ and for

$$s = \frac{1}{\sum_{i \in N} u_i(x_i)} U(x)$$

the value $f(s)s$ is attained by $U(\hat{X})$. However, with the hypotheses so far there is no reason for the value $f(s)$ to be attained for every s, or equivalently for $U(\hat{X})$ to be closed. In other words, while weakly efficient allocations exist trivially (just give the entire ω to any consumer), interesting ones, e.g., with some prescribed sharing of "utility," are not guaranteed to exist.

Example 2 (Araujo (1985)): Let $L = l_\infty$ be the space of bounded infinite sequences endowed with the sup norm. Take $\omega = (1, \dots, 1, \dots)$ and $N = 2$. The preferences of $i = 1$ (resp. $i = 2$) are given by $x \succsim_i y$ if and only if $\lim_{n\to\infty} \inf x_n \geqslant \lim_{n\to\infty} \inf y_n$ (resp. $\sum_{n=1}^{\infty} (1/2^n)x_n \geqslant \sum_{n=1}^{\infty} (1/2^n)y_n$). It is simply seen that the set $U(\hat{X})$ has then the form indicated in Figure 3 (the point (0, 1) is included). Note that the first consumer is absolutely patient while the second discounts the future.

As emphasized by Araujo (1985) we can hardly expect to prove the existence of equilibrium prices in our (finite number of consumers) context if our model does not guarantee in any generality the existence of weakly efficient allocations. What is needed is that we be able to replace sup. by max. in the definition of $f: \Delta \to [0, N]$ or, equivalently in view of Proposition 2, that $U(\hat{X})$ be closed. This is what we shall assume (we give the condition directly in terms of preferences).

Closedness Hypothesis: Consider a sequence $x_n \in \hat{X}$ and suppose that $n > m$ implies $x_{ni} \succsim_i x_{mi}$ for all i. Then there is $x \in \hat{X}$ such that $x_i \succsim_i x_{ni}$ for all n and $i \in N$.

It is simply seen that the above hypothesis is equivalent to the closedness of $U(\hat{X})$. Of course, the Closedness Hypothesis would not be very good if one could not find conditions on individual preferences which yield it. But this is easy. Let the sequence x_n be as in the hypothesis. Then $V_n = \bigcap_{i\in N} \{x \in X^N : x_i \succsim_i x_{ni}\} \cap \hat{X}$ constitutes a nested sequence of closed sets. If $\bigcap_n V_n$ is nonempty then we are done. This intersection will be nonempty if in a topology for which the convex sets $\{z \in X : z \succsim_i y\}$, $i \in N$, $y \in X$, are

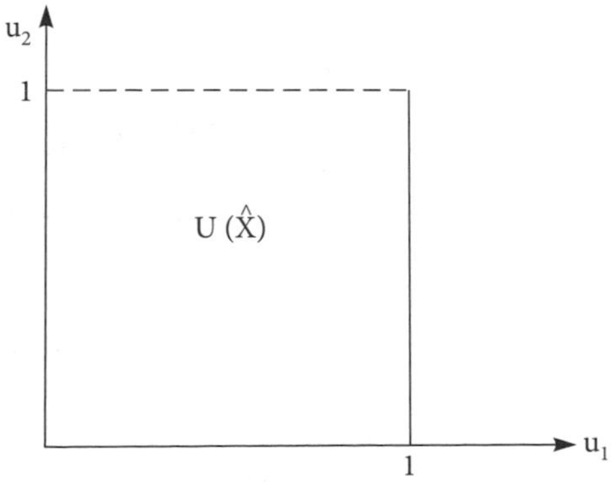

Figure 3

closed, the set $\hat{X}$, which is convex, bounded below by 0 and above by ω, is compact. If L is dual to some other space B then $\hat{X}$ is compact in the weak-star topology (i.e., the topology of the pointwise convergence when members of L are looked at as linear functional on B). For reflexive spaces such as L_p, $1 < p < \infty$, the weak-star coincides with the weak topology. Because sets which are norm closed and convex are weakly closed, the closedness hypothesis is automatically satisfied on them. For L_∞ and $ca(K)$, which have preduals but are not reflexive, the hypothesis that the sets $\{z \in X : z \succsim_i y\}$ are weak-star closed is strong but it has clear and natural economic interpretations. In L_∞ it means that preferences "discount" the future (see Bewley (1972), and Brown-Lewis (1982)) while in $ca(K)$ it says that commodity bundles with similar characteristics are treated similarly by preferences (see Mas-Colell (1975), and Jones (1984a)). The space L_1 has no predual but it is nonetheless still true that $\hat{X}$ is weakly compact (because it is order-bounded; I owe this observation to C. Aliprantis).

7. Supporting Prices for Weakly Efficient Allocations

This is the key section of this paper. Suppose that individual preferences are continuous, monotone, and convex and that x is a weakly efficient allocation. We will look for conditions guaranteeing the existence of a supporting price functional, i.e., a $p \in L^*_+$, $p \neq 0$, such that $z \succsim_i x_i$ implies $p \cdot z \geqslant p \cdot x_i$ for every i. In the next section we shall see that once these supporting prices are available, the rest of the existence proof is fairly routine sailing.

Obviously, we shall have to require that the preferences of individual consumers be proper. As we saw in Section 4, properness of $\succsim_i$ at x_i is equivalent to the existence of $p_i \in L^*_+$, $p_i \neq 0$, such that $z \succsim_i x_i$ implies $p \cdot z \geqslant p \cdot x_i$. One may have hoped that, with x a weak optimum, properness, or at least uniform properness, would imply social supportability (i.e., "social properness"). Unfortunately, this does not follow automatically as the following example, due to Jones (1984b), demonstrates. Experts in urban economics will have no difficulty in recognizing it.

Example 3 (Jones): Let $L = L_\infty([0,1])$ have the topology induced by the duality with $C^1([0,1])$, i.e., $y_n \to y$ if and only if $\int_0^1 y_n(t)g(t)\,dt \to \int_0^1 y(t)g(t)\,dt$ for every continuously differentiable g. Let $N = 2$ and the preferences of $i = 1$ (resp. $i = 2$) be represented by $u_1(x_1) = \int_0^1 (1-t)\, x_1(t)\, dt$ (resp. $\int_0^1 t x_2(t)\, dt$). The utility functions are linear and continuous. Hence, preferences are uniformly proper. Put $\omega(t) = 1$ for all t. The feasible allocation x defined by $x_1(t) = 1$ if $t \leqslant \frac{1}{2}$, $x_1(t) = 0$ if $t \geqslant \frac{1}{2}$ (resp. $x_2(t) = 0$ if $t \leqslant \frac{1}{2}$, $x_2(t) = 1$ if $t \geqslant \frac{1}{2}$) is weakly optimal. However, it cannot be supported by any continuous linear functional since the only supporting functional, which is given by $q(t) = 1 - t$ if $t \leqslant \frac{1}{2}$, $q(t) = t$ if $t \geqslant \frac{1}{2}$, cannot be expressed as a continuously differentiable function of t and, therefore, it is not continuous as a linear functional on L.

The main result of this section is that if the commodity space has the structure of a *topological vector lattice* then individual uniform properness implies social supportability.

An ordered linear space is a *vector lattice* (or Riesz space) if the order $\geqslant$ of the space is a lattice order, i.e., for any $x, y \in L$ there are elements of the space, denoted $x \vee y$, $x \wedge y$, such that $z \geqslant x$, $z \geqslant y$ (resp. $z \leqslant x$, $z \leqslant y$) implies $z \geqslant x \vee y$ (resp. $z \leqslant x \wedge y$). For any $x \in L$ one can then define the positive part $x^+ = x \vee 0$, the negative part $x^- = (-x) \vee 0$, and the absolute value $|x| = x^+ + x^-$.

If, in addition, the order and topological structures of the space fit together well enough for the lattice operations to be uniformly continuous, then we have a *topological vector lattice* (or locally solid Riesz space). A salient and all important example is the *Banach lattices,* which are vector lattices admitting a complete norm $\|\ \|$ with the property that $|x| \leqslant |y|$ implies $\|x\| \leqslant \|y\|$. The commodity space of Example 3 fails to be a topological vector lattice.

Although positive cones inducing lattice orders are mathematically special (for example, if $L = R^n$ then L_+ is a lattice cone if and only if it is generated by precisely n extreme rays), economic examples (including Example 3 above) fall naturally into a vector lattice framework. We do not view, therefore, this part of the hypothesis as restrictive. It is quite another matter with the requirement that the lattice operations be continuous. It is here that the strength of the hypothesis lies. We shall defer a more detailed discussion of this point until after the next Proposition.

Proposition 3: Suppose that L is a topological vector lattice and that the preference relations $\succsim_i$, $i \in N$, on L_+ are convex, monotone and uniformly proper. For each $i \in N$ let $v_i \in L_+$ be as in the definition of properness. Put $v = v_1 + \cdots + v_N$. Then there is $V \subset L$, an open neighborhood of zero, such that any weakly optimal allocation can be supported by a $p \in L_+^*$ with $p \cdot v = 1$ and $|p \cdot z| \leqslant 1$ for all $z \in V$.

Proof: For every i let $V_i \subset L$ be the neighborhood of zero given in the definition of uniform properness. Without loss of generality we can assume that V_i is convex (the space L is locally convex) and that $V_i = -V_i$. Put $V = V_1 \cap \cdots \cap V_N$ and let $\Gamma \subset L$ be the open, convex cone spanned by $\{v\} + V$.

For a given weakly optimal allocation x define

$$Z = \left\{ \sum_{i=1}^{N} (z_i - x_i) : z \succsim_i x_i \right\} \subset L.$$

From this point on, the proof proceeds in two steps. The first shall establish the collective properness property, namely, that for an adequately chosen Γ we have $Z \cap (-\Gamma) = \varnothing$. The proof does not require the convexity of preferences. It relies crucially on the decomposition property for vector lattices and on the (uniform) continuity of the lattice operations. So the full strength of the topological vector lattice assumption is used in the first step. The second step is then a straightforward separation argument relying on the Hahn-Banach theorem; convexity of preferences is, of course, essential here.

Step 1: Because the lattice operations are uniformly continuous, we can choose the sets V_i so that they are *solid,* i.e., $|u| \leqslant |z|$, $z \in V_i$, implies $u \in V_i$ (see Aliprantis and Burkinshaw (1978, p. 34)). To show that $Z \cap -\Gamma = \varnothing$ we argue by contradiction. Suppose there are $z_i \succsim_i x_i$ such that, denoting $\omega = \sum_{i=1}^{N} x_i$ and $z = \sum_{i=1}^{N} z_i$, we have $z - \omega \in -\Gamma$, i.e., $z - (\omega - \alpha v) \in \alpha V$ for some $\alpha > 0$. Call $y = \omega - \alpha v$. We claim that $(y - z)^{-} \leqslant z + \alpha v$. This can be seen as follows. Because $z \geqslant 0$ we have $y - z \leqslant y \leqslant \omega$. Hence, $(y - z)^{+} = (y - z) \vee 0 \leqslant \omega$. So,

$$z = y - \omega + \omega - (y - z)^{+} + (y - z)^{-} \geqslant -\alpha v + (y - z)^{-}.$$

Because $z + \alpha v = \Sigma_{i \in N} (z_i + \alpha v_i)$ the Decomposition Property of vector lattices (see, for example, Aliprantis and Burkinshaw (1978, p. 3)) implies that we can write $(y - z)^{-} = \Sigma_{i \in N} s_i$ where $0 \leqslant s_i \leqslant z_i + \alpha v_i$ for each $i \in N$. Define now $z_i' = z_i + \alpha v_i - x_i \geqslant 0$. Suppose that, for some i, $z_i \succsim_i z_i'$. Because of the properness property this implies $s_i \notin \alpha V$. On the other hand, $0 \leqslant s_i \leqslant (y - z)^{-} \leqslant |y - z|$ and $y - z \in \alpha V$. Hence, $s_i \in \alpha V$ and we have a contradiction. Therefore, $z_i' \succ_i z_i \succsim_i x_i$ for every i. But this contradicts the weak efficiency of x because

$$\sum_{i=1}^{N} z_i' = z + \alpha v - (y - z)^- \leqslant z + \alpha v - (y - z)^- + (y - z)^+$$
$$= z + \alpha v + (y - z) = y + \alpha v = \omega = \sum_{i=1}^{N} x_i.$$

Step 2: Because of the convexity hypothesis on preferences, the open set $Z + \Gamma$ is convex. Also, $0 \notin Z + \Gamma$ because $Z \cap -\Gamma = \emptyset$. Therefore, by the Hahn-Banach Theorem (see Schaefer, 1971, p. 46) there is a nonzero, continuous linear functional p such that $p \cdot y > 0$ for all $y \in Z + \Gamma$. Remembering that Γ is a cone and $0 \in Z$ this yields $p \cdot z \geqslant 0$ for all $z \in Z$ and $p \cdot y > 0$ for all $y \in \Gamma$. In particular, $p \cdot v > 0$. So, without loss of generality, we can take $p \cdot v = 1$. If $z \in V$ then $v - z \in \Gamma$ and so $p \cdot z \leqslant p \cdot v = 1$. Since the same applies to $-z$ we get $|p \cdot z| \leqslant 1$. Finally, let $z \succsim_i x_i$ Then $z - x_i \in Z$ and so, $p \cdot z \geqslant p \cdot x_i$. We conclude that p supports x. The fact that $p \geqslant 0$ follows from the monotonicity of any $\succsim_i$. *Q.E.D.*

The requirement that the lattice operations be uniformly continuous is not minor. It places strong restrictions on the topology of L. For example, if L is a Banach lattice then the lattice operations are uniformly continuous in the weak topology if and only if the space is finite dimensional (Aliprantis and Burkinshaw (1978, p. 42)). Of course, if L is a Banach lattice then Proposition 3 gives us in all generality a price functional in the norm dual. If L is reflexive then this is precisely what one wants. But if as L_∞ and $ca(K)$ the space L has a predual $\hat{L}$ strictly smaller than its dual L^* then we may ideally want a price functional in $\hat{L}$, i.e., in L_1 for L_∞ and in $C(K)$ for $ca(K)$. One way to aim at this is to work with a topology on L generating $\hat{L}$ as dual. It is by no means guaranteed that such a topology, weaker than the norm, will preserve the continuity of the lattice operations. Consider, for example, the sequence $x_n = \delta_{(1/n)} - \delta_0$ in $ca([0, 1])$; it converges to 0 in the weak-star topology but $x_n^+ \to \delta_0 \neq 0$. The space L_∞ is again better behaved: the Mackey topology induced by its pre-dual does the trick (I owe this observation to C. Aliprantis and W. Zame). In spite of all this, Proposition 3 is still quite relevant because by using a price system in the norm dual it is often possible to strengthen the properties of the price functional through space-specific arguments.

We end this section with four remarks:

(i) One feels there is still considerable room for improvement between the level of pathology of Example 3 and the strength of the continuity hypothesis of Proposition 3. See Mas-Colell (1986) for an exploration of this gap.

(ii) Even to establish supportability at a single weak optimum, the proof of Proposition 3 makes essential use of uniform properness as opposed to properness at a point.

(iii) Over and above the commodity space being a topological vector lattice, it is also important for Proposition 3 that the consumption set be the positive orthant (or at least an order-closed subcone of the positive orthant containing v_i). Relaxing this would be of interest, mostly for the insights it would provide into the generalization to the production case.

(iv) It is to be noted that the lattice theoretic hypothesis on the space and the consumption set cannot be dispensed with even in the finite-dimensional case. While the existence of a collective supporting price p follows purely from convexity considerations, this is not so for the requirement $p \cdot v = 1$. For an investigation of the finite-dimensional case see Yannelis-Zame (1984, Sec. 7) and Mas-Colell (1985a).

8. An Equilibrium Existence Theorem

All the ingredients are now in place for the existence proof.

Suppose that $\omega = \Sigma_{i \in N} \omega_i$, $\omega_i \geqslant 0$, i.e., every consumer has a specific claim on a part of the initial endowment vector. A feasible allocation $x \in L_+^N$ and a $p \in L_+^*$, $p \neq 0$, constitute a *quasiequilibrium* if, for all i, $p \cdot \omega_i = p \cdot x_i$ and $p \cdot z \geqslant p \cdot x_i$ whenever $z \succsim_i x_i$. An *equilibrium* if in fact $z \succsim_i x_i$ implies $p \cdot z > p \cdot x_i$ *for* all i. The latter property holds at a quasiequilibrium for any i such that $p \cdot \omega_i > 0$.

Theorem: Let L be a topological vector lattice and $\succsim_i$, $i \in N$, *a collection of monotone, convex, continuous and uniformly proper preference relations on the positive orthant of L. Suppose that for the economy* $\{(\succsim_i, \omega_i)\}_{i \in N}$ *the vector* $\omega = \sum_{i=1}^N \omega_i$ *is desirable for every* $i \in N$ *and that the Closedness Condition holds. Then there is a quasiequilibrium* (x, p). *Moreover,* $p \cdot v = 1$ *where* $v = v_1 + \cdots + v_N$ *and* v_i *is the properness vector for* $\succsim_i$.

Proof: The demonstration is a straightforward adaptation (in fact, almost a transcription) of Bewley's (1969) and Magill's (1981) proof for the case L_∞. It consists of three steps. In the first we define a certain correspondence $\Phi : \Delta \to T = \{z \in R^N : \Sigma_{i \in N} z_i = 0\}$ which has the property that any of its zeroes yields a quasiequilibrium. In the second we show that Φ is convex valued and upper-hemicontinuous. Finally in the third we prove that Φ has a zero.

Step 1: Take $f\colon \Delta \to [0, N]$ as in Section 6 and $v \in L_+$, $V \subset L$, as in Proposition 3. For any $s \in \Delta$ pick a feasible allocation $x(s)$ such that $U(x(s)) = f(s)s$. We can, without loss of generality, assume that $\Sigma_{i \in N}\, x_i(s) = \omega$ (indeed, not everybody can be made better off, which means that if there is any surplus left it can be distributed so as to make no one better off). Then let $P(s) = \{p \in L^*_+ \colon p \cdot v = 1, |p \cdot z| \leqslant 1$ for all $z \in V$, p supports the weakly efficient allocation $x(s)\}$. Obviously, $P(s)$ is convex and, by Proposition 3, nonempty. Finally, put $\Phi(s) = \{(p \cdot (\omega_i - x_i(s)), \ldots, p \cdot (\omega_N - x_N(s)))\colon p \in P(s)\}$. Then $\Phi(s)$ is non-empty and convex valued, $\Sigma_{i \in N}\, z_i = 0$ for any $z \in \Phi(s)$, and $0 \in \Phi(s)$ if and only if $x(s)$ is a quasiequilibrium.

Step 2: This is the key step. We establish that $\Phi : \Delta \to T$ is an upper hemicontinuous correspondence. Let $s_n \to s$, $w_n \in \Phi(s_n)$, $w_n \to w \in T$. We should show that $w \in \Phi(s)$. Suppose that $p_n \in P(s_n)$ is the price vector yielding w_n. Because $|p_n \cdot z| \leqslant k$ for all $z \in V$ we can assume, by Alaouglou's Theorem (see Schaefer (1971, p. 84)), that p_n has a weak-star limit, i.e., $p_n \cdot y \to p \cdot y$ for any $y \in L$. This yields $p \cdot v = 1$. Also, let $z \succsim_i x_i(s)$. Because $f(s_n)s_n \to f(s)s$ we should eventually have $z \succsim_i x_i(s_n)$ which implies $p_n \cdot z \geqslant p_n \cdot x_i(s_n) = p_n \cdot \omega_i - w_{ni}$. Taking limits, $p \cdot z \geqslant p \cdot \omega_i - w_i$. Because of monotonicity we can conclude that $p \cdot z \geqslant p \cdot \omega_i - w_i$ whenever $z \succsim_i x_i(s)$. In particular, $p \cdot x_i(s) \geqslant p \cdot \omega_i - w_i$ for all i. Since $\Sigma_{i \in N}\, x_i(s) = \Sigma_{i \in N}\, \omega_i$ we in fact have $p \cdot x_i(s) = p \cdot \omega_i - w_i$. This proves that $w \in \Phi(s)$.

Step 3: The proof that $\Phi : \Delta \to T$ has a zero is routine. We simply consider the upper hemicontinuous, nonempty, convex valued map $s \mapsto s + \Phi(s)$ on Δ and note that it satisfies suitable boundary conditions for an application of Kakutani's fixed point theorem. Indeed, $s_i = 0$ and $w \in \Phi(s)$ implies $w_i \geqslant 0$ because $u_i(x_i(s)) = 0$, i.e., $0 \succsim_i x_i(s)$, and so $0 \geqslant p \cdot x_i(s) \geqslant 0$ for any p supporting $x(s)$. *Q.E.D.*

The Theorem would not be of much interest if we could have $p \cdot \omega = 0$. The way to avoid this sort of degeneracy is to make the natural assumption that, for all $i \in N$, ω is desirable in the (relatively) strong sense of being able to take $v = \omega$. Then the Theorem guarantees $p \cdot \omega = 1$. This, incidentally, is not an issue specific to the infinitely many commodities context. It can arise with one consumer and two commodities.

Finally, a comment on the initial motivation for this research, the space $ca(K)$. The Theorem makes clear that the bulk of smoothness hypotheses of Jones (1984a) is needed to get the price system in $C(K)$. Only a minor part of them (roughly, the part that does not depend on the topological structure of K) is required for the preferences to be proper with respect to the norm topology and to get, therefore, a price system in the norm dual.

9. Extensions and Related Recent Work

Yannelis and Zame (1984) have reproved and extended (to unordered, locally nonsatiated preferences) the Theorem of Section 8 using a technique of approximation by finite-dimensional economies. In a more limited context, a different proof of the same general nature is outlined in Brown (1983). An enlightening elucidation of the role of the uniform properness condition is contained in Aliprantis-Brown-Burkinshaw (1985).

Our Theorem depends on the consumption sets being the positive orthant of the space. This suggests that the extension to production economies is a nontrivial task. Duffie-Huang (1983b), Duffie (1984), and Jones (1984c) have provided generalizations to production. All assume some form of noninteriority of the production set. With this (strong) hypothesis the theory is quite parallel to the L_∞ case. Properness assumptions are dispensable and weak optima are supportable (this was proved by Debreu (1954a)). The existence theorems of Aliprantis-Brown (1983), Yannelis (1984), and Simmons (1985), where excess demand functions are primitives, could also be interpreted in terms of noninteriority assumptions on production sets. A model where the consumption rather than the production sets have nonempty interior, is in Chichilnisky-Heal (1984); see also Kreps (1981). Recently, Zame (1985) has succeeded in finding an ingenious way to exploit lattice theoretic properties in order to obtain a Theorem which does incorporate nontrivial production while avoiding a nonempty interior hypothesis. See also Mas-Colell (1986).

Another topic for further research, the weakening of the continuity hypothesis on the lattice operations, has already been mentioned in Section 7.

Our method of proof is so dependent on the finiteness of the number of agents that it prompts one to ask how the double infinity could be handled in an abstract setting. An important reference is Ostroy (1984) where a result for the commodity space L_1 is obtained. The problem tackled in Section 7 does not arise in Ostroy's work because, in effect, the properness-like hypothesis is imposed directly at the collective level. This is, of course, well in accord with the dictum that it is best to face problems one at a time.

Harvard University, Littauer Center, Cambridge, MA 02138, U.S.A.

Manuscript received October, 1983; *final revision received July,* 1985.

References

Aliprantis, C., and D. Brown (1983): "Equilibria in Markets with a Riesz Space of Commodities," *Journal of Mathematical Economics,* 11, 189–207.

Aliprantis, C., D. Brown, and O. Burkinshaw (1983): "Edgeworth Equilibria," Yale University, Cowles Discussion Paper No. 756.

Aliprantis, C., and O. Burkinshaw (1978): *Locally Solid Riesz Spaces.* New York: Academic Press.

Araujo, A. (1985): "Lack of Equilibria in Economies with Infinitely Many Commodities: the Need of Impatience," *Econometrica,* 53, 455–462.

Bewley, T. (1969): "A Theorem on the Existence of Competitive Equilibria in a Market with a Finite Number of Agents and Whose Commodity Space is L_∞," CORE Discussion Paper, Université de Louvain.

——— (1972): "Existence of Equilibria in Economies with Infinitely Many Commodities," *Journal of Economic Theory,* 4, 514–540.

Bojan, P. (1974): "A Generalization of Theorems on the Existence of Competitive Economic Equilibrium to the Case of Infinitely Many Commodities," *Mathematica Balkanika,* 4, 490–494.

Brown, D. (1983): "Existence of Equilibria in a Banach Lattice with an Order Continuous Norm," Yale University, Cowles Preliminary Paper No. 91283.

Brown, D., and L. Lewis (1981): "Myopic Economic Agents," *Econometrica,* 49, 359–368.

Chamberlain, G., and M. Rothschild (1983): "Arbitrage, Factor Structure, and Mean-Variance Analysis on Large Asset Markets," *Econometrica,* 51, 1281–1304.

Chichilnisky, G., and G. Heal (1984): "Competitive Equilibrium in L_p and Hilbert Spaces with Unbounded Short Sales," Columbia University, mimeographed.

Chichilnisky, G., and P. Kalman (1980): "An Application of Functional Analysis to Models of Efficient Allocation of Resources," *Journal of Optimization Theory and Applications,* 30, 19–32.

Debreu, G. (1954a): "Valuation Equilibrium and Pareto Optimum," *Proceedings of the National Academy of Sciences,* 40, 588–592.

——— (1954b): "Representation of a Preference Ordering by a Numerical Function," *Decision Processes*, ed. by R. Thrall, C. Coombs, and R. Davis. New York: J. Wiley.

——— (1962): "New Concepts and Techniques for Equilibrium Analysis," *International Economic Review*, 3, 257–273.

——— (1982): "Existence of Competitive Equilibrium," in *Handbook of Mathematical Economics*, Vol. II, ed. by K. Arrow and M. Intriligator. Amsterdam: North-Holland.

Duffie, D. (1983): "Competitive Equilibria in General Choice Spaces," mimeographed, Graduate School of Business, Stanford University, forthcoming in the *Journal of Mathematical Economics*.

Duffie, D., and C. Huang (1983a): "Implementing Arrow-Debreu Equilibria by Continuous Trading of Few Long-lived Securities," mimeographed, Stanford University.

——— (1983b): "Competitive Equilibria with Production in Infinite Dimensional Commodity Spaces," mimeographed, Graduate School of Business, Stanford University.

El'Barkuki, R. A. (1977): "The Existence of an Equilibrium in Economic Structures with a Banach Space of Commodities," *Akad. Nauk. Azerbaidjan*, SSR Dokl. 33, 5, 8–12 (in Russian with English summary).

Florenzano, M. (1983): "On the Existence of Equilibria in Economies with an Infinite Dimensional Commodity Space," *Journal of Mathematical Economics*, 12, 270–219.

Harrison, M., and D. Kreps (1979): "Martingales and Arbitrage in Multiperiod Securities Markets," *Journal of Economic Theory*, 20, 381–408.

Jones, L. (1984a): "A Competitive Model of Product Differentiation," *Econometrica*, 52, 507–530.

——— (1984b): "Special Problems Arising in the Study of Economies with Infinitely Many Commodities," MEDS Discussion Paper No. 596, Northwestern Univesity.

——— (1984c): "A Note on the Price Equilibrium Existence Problem in Banach Lattices," MEDS Discussion Paper No. 600, Northwestern University.

Kannai, Y. (1970): "Continuity Properties of the Core of a Market," *Econometrica*, 38, 791–815.

Khan, M. Ali (1984): "A Remark on the Existence of Equilibria in Markets Without Ordered Preferences and With a Riesz Space of Commodities," 13, 165–171.

Kreps, D. (1981): "Arbitrage and Equilibrium in Economies with Infinitely Many Commodities," *Journal of Mathematical Economics*, 8, 15–36.

Magill, M. (1981): "An Equilibrium Existence Theorem," *Journal of Mathematical Analysis and Applications*, 84, 162–169.

Mas-Colell, A. (1975): "A Model of Equilibrium with Differentiated Commodities," *Journal of Mathematical Economics,* 2, 263–296.

——— (1985a): "Pareto Optima and Equilibria: the Finite Dimensional Case," in *Advances in Equilibrium Theory,* ed. by C. Aliprantis, O. Burkinshaw, and N. Rothman. New York: Springer-Verlag, pp. 25–42.

——— (1986): "Valuation Equilibrium and Pareto Optimum Revisited," in *Contributions to Mathematical Economics,* W. Hildenbrand and A. Mas-Colell, eds. Amsterdam: North-Holland.

Monteiro, P. (1985): "Some Results on the Existence of Utility Functions on Path Connected Spaces," IMPA, Rio de Janeiro, mimeographed.

Negishi, T. (1960): "Welfare Economics and Existence of an Equilibrium for a Competitive Economy," *Metroeconomica,* 12, 92–97.

Ostroy, J. (1984): "On the Existence of Walrasian Equilibrium in Large-Square Economies," *Journal of Mathematical Economics,* 13, 143–164.

Richards, S. F. (1985a): "Prices in Banach Lattices with Concave Utilities," Carnegie-Mellon University, mimeographed.

——— (1985b): "Prices in Banach Lattices with Convex Preferences," Carnegie-Mellon University, mimeographed.

Schaeffer, H. (1971): *Topological Vector Spaces.* New York: Springer-Verlag.

——— (1974): *Banach Lattices and Positive Operators.* New York: Springer-Verlag.

Shafer, W. (1984): "Representation of Preorders on Normed Spaces," University of Southern California, mimeographed.

Simmons, S. (1985): "Minimaximin Results with Applications to Economic Equilibrium," *Journal of Mathematical Economics,* 13, 289–304.

Toussaint, S. (1984): "On the Existence of Equilibria in Economies With Infinitely Many Commodities," *Journal of Economic Theory,* 33, 98–115.

Yannelis, N. (1984): "On a Market Equilibrium Theorem With an Infinite Number of Commodities," University of Minnesota, mimeographed.

Yannelis, N., and N. D. Prabhakar (1983): "Existence of Maximal Elements and Equilibria in Linear Topological Spaces," *Journal of Mathematical Economics,* 12, 233–245.

Yannelis, N., and W. Zame (1984): "Equilibria in Banach Lattices Without Ordered Preferences," Institute for Mathematics and its Applications, Preprint #71, University of Minnesota.

Zame, W. (1985): "Equilibria in Production Economies with an Infinite Dimensional Commodity Space," Institute for Mathematics and its Applications, Preprint No. 127.

— 6 —

Real Indeterminacy with Financial Assets*

JOHN GEANAKOPLOS
ANDREU MAS-COLELL

It is shown that in a two period general equilibrium securities model where assets pay in money the generic dimension of the set of equilibrium allocations, in the incomplete market situation, is $S - 1$, where S is the number of assets. Hence the degree of real indeterminacy is independent of the number of assets. This result requires, beyond fewer assets than states, that the number of traders be larger than the number of securities and that the asset return matrix be in general position. The generic dimension for arbitrary returns matrix is also obtained. It is argued, in addition, that the presence of real or mixed assets does not by itself lower the degree of indeterminacy. *Journal of Economic Literature* Classification Numbers: 021, 022, 023.

I. Introduction

It has long been known in economics that the notion of general competitive equilibrium displays a basic multiplicity, though this indeterminacy has usually been disposed of as being almost entirely nominal. An Arrow-Debreu economy, for example, typically has a continuum of equilibrium price vectors, but only a finite number of these give rise to distinct com-

Journal of Economic Theory 47 (1989), 22–38.

* We have benefited at different stages by comments by many people, among them D. Cass, D. Duffie, J. Moore, A. Roëll, J. Stiglitz, S. Werlang, and an anonymous referee. Our very special debt to D. Cass should be clear from the paper. Y. Balasko and D. Cass have independently undertaken a study similar to ours. Theorem 1′ in Section III has been obtained jointly with J. Moore. Financial support from NSF is gratefully acknowledged. The second author also wishes to thank MSRI, at Berkeley, The Department of Economics at Berkeley, and the Guggenheim Foundation.

modity allocations. The accounting relation called Walras Law implies that the economy-wide system of excess demands has one more endogenous price than it has independent market clearing equations. However, the homogeneity of demand, i.e., the fact that the aggregate excess demand function depends on relative prices and not on their absolute level, explains why most of the ensuing indeterminacy is nominal. Our starting point here is the observation that for economics less idealized than that of Arrow-Debreu, involving the exchange of monetarized assets, the indeterminacy caused by Walras Law is greater than one-dimensional, and because there is no corresponding increase in the homogeneity of demand, the difference manifests itself as a real indeterminacy of equilibrium.

In this paper we draw a sharp distinction between economies in which the assets promise delivery in a money (say green pieces of paper) whose exchange value can exceed its (marginal) use value, and those economies where the assets deliver in a commodity money whose exchange value is tied to its use value. In the latter situation the lock-step balance between Walras Law and the homogeneity of excess demand preserves local uniqueness of real equilibrium. But in the money case there is usually a multidimensional continuum of competitive equilibria, each representing a different commodity allocation. Often there will be Pareto comparability between equilibria.

An important, preliminary, example of real indeterminacy in a "monetary" economy occurs in the standard static Walrasian setting if we add an extra commodity, which we call money, that has no effect on any agent's utility. Let each agent be endowed with m^h green dollar bills (i.e., units of the money). As long as $\Sigma_h\ m^h > 0$, we know that the equilibrium price of money must be equal to its use value, namely zero. But if $\Sigma_h\ m^h = 0$, and we allow m^h to be negative or positive, then the price of money is not tied to zero, and in fact it is easy to show that typically there is a one-dimensional continuum of equilibria involving different commodity allocations. The same is true if $\Sigma_h\ m^h > 0$, but each agent h owes a money tax d^h to the government with $\Sigma_h\ d^h = \Sigma_h\ m^h$.[1]

The reason for this real indeterminacy with money can be expressed in two equivalent ways. Note that the excess demand for money is degenerate, i.e., at any vector of prices the demand for money will match its supply.

1. An old argument, usually attributed to Abba Lerner, holds that government money taxes are a major reason government issued paper money has value.

Introducing money thus adds one more variable price, but does not add another independent market clearing condition. Equivalently, if the price of each dollar is fixed at 1, demand for real commodities is no longer homogeneous in commodity prices, yet Walras Law still applies to the commodities, so that if all the commodity markets clear but one, this last will clear as well. Note that both explanations of indeterminacy rest on the fact that at least for some h, $m^h \neq d^h$; i.e., the economy is in midstream with some agents already debtors and others creditors.[2] Otherwise commodity demand would be homogeneous in commodity prices (or equivalently, the price of money would not affect commodity demand). From now on we shall always include money in our models, and we shall always fix the price of each dollar at 1, ignoring equilibria in which money is valueless. All prices are thus quoted in dollars, which become the units of account. When we use the terms Walras Law and homogeneity we shall mean with regard to commodity prices and demands, holding the money price fixed at 1.

Consider now an economy in which trade takes place sequentially, perhaps in different states of nature. Let trade take place in period zero ($s = 0$) and again in period 1 for each state $s = 1, \ldots, S$. Assume that every consumer's endowment of money in every state is zero. In each state s each agent is required to balance the value of his expenditures and sales.[3] Agents therefore face $S + 1$ budget constraints. Since Walras Law can be applied $S + 1$ times, there are $S + 1$ redundant market clearing equations and we should expect $S + 1$ dimensions of indeterminacy of equilibrium prices. However, it is clear that there are also $S + 1$ independent applications of homogeneity, since each state's commodity prices can be scaled independently without affecting demand. It is in fact easy to show that typically there are only a finite number of distinct real equilibria. Thus although

2. Irving Fisher [8] recognized very clearly the indeterminacy of equilibrium in a monetary economy caused by unanticipated (mid-stream) fluctuations in the absolute commodity price level. Indeed he advocated a government engineered inflation as a way of transferring wealth from creditors to debtors in order to pull the American economy out of the Great Depression. We shall see in a moment that even perfectly anticipated price level changes can have real effects when there is uncertainty and incomplete asset markets.

3. One interpretation of our model, in which money appears as the medium of exchange, is as follows. Agents are able to borrow as much money (dollar bills) from the central bank as they wish at the beginning of each state. Agents can then buy goods with their money or sell goods for money. At the end of the state-period they must pay back to the central bank exactly as many dollars as they have borrowed.

there are $S + 1$ dimensions of equilibrium price vectors, differing by their absolute levels across the states, most of this indeterminacy has no effect on real consumption.

Let us next enrich the model along the lines suggested by Radner [13] and Hart [11] by allowing agents to trade in period 0 prespecified real assets as well as commodities. A real asset is a promise to deliver a vector of state contingent commodity bundles; agents can be allowed to buy or sell these claims. If every conceivable real asset were traded, the model would reduce to a special case of the Arrow-Debreu model. Whether or not the real asset market is complete, it is evident that since the assets are prefectly "indexed," there are again $S + 1$ independent operations of homogeneity, and it can generally be shown that generic, real asset economies have a finite number of distinct real equilibria (see Geanakoplos-Polemarchakis [9]).

The first general equilibrium model involving assets occurs in Arrow [1] where assets are promises to deliver state contingent dollars. To distinguish these monetary assets from the real assets above, we shall call them financial assets or securities. Arrow concentrated on a specialized type of financial asset that promises delivery of one dollar in precisely one state s, and nothing in the other states. In his honor these have come to be called Arrow securities. Arrow proved the remarkable result that when agents are permitted to trade all S Arrow securities and spot prices are correctly anticipated then the equilibrium commodity allocations are identical to those that would arise in the Arrow Debreu model discussed above where agents are permitted to trade all possible state contingent commodities. We can conclude that typically, in a complete Arrow securities economy almost all the indeterminacy is nominal (i.e., not real). Again Arrow's result may be looked at as a balance between the $S + 1$ occurrences of Walras Law and of the homogeneity of demand, but the homogeneity is more subtle than before. As usual, demand for assets and commodities is in period zero homogeneous of degree zero relative to period zero prices. For each state $s \geqslant 1$, the demand for the sth Arrow security is homogeneous of degree one,[4] and the demand for all the other securities and commodities is homogeneous of degree zero, in the absolute level of prices in state s, provided that the asset price for the sth Arrow security is varied inversely with p_s. Once a

4. Since the initial endowment of assets is zero for every agent, homogeneity of any degree is enough to negate one dimension of multiplicity.

state s is realized, asset promises will make some agents creditors and others debtors. According to our preliminary example, changes in the absolute level of prices p_s can have real effects. What happens is that if these price changes are anticipated, then rational agents will readjust their portfolios of Arrow securities so that in the end there are no real effects.

In this paper we consider economies with an incomplete set of arbitrary financial assets, as in Cass [4], Werner [14], and Duffie [7]. These papers all suggest that there may be real indeterminacy. In fact, Cass [3] constructs an explicit example with one financial asset and two states in which there is a one-dimensional continuum of distinct real equilibria. In this paper we follow Cass' lead by taking up the general problem of real indeterminacy with financial assets. We find that "typically," any change in the relative rates of inflation (from 0 to s) across the states has a real effect, even if it is perfectly anticipated. This means that there are $S - 1$ degrees of real indeterminacy. It is clear that there are at least two independent sources of homogeneity in demand, including the usual homogeneity in period zero asset and commodity prices. The second source comes from the fact that if all period one prices are doubled, and all asset prices halved, then commodity demand is unaffected and asset demands are doubled. It is not clear whether there are other sources of homogeneity, and our result implies there cannot be. The occurrences of Walras Law provide no more than $S + 1$ degrees of freedom. Of these we see that $S - 1$ are real and 2 are nominal.

More precisely, we begin by fixing the smooth preferences of each agent and the state contingent dollar payoffs of each financial asset. The payoffs of the assets can be summarized by an $S \times B$ matrix R. We say that the asset payoffs are in general position if every submatrix of R is of full rank. Clearly if the payoffs were chosen randomly, R would nearly always be in general position. Our main result is given by Theorem 1, in Section II, which essentially asserts that if there are fewer assets than states ($B < S$), i.e., if the asset market is incomplete, and if the assets are in general position, then, provided there are at least as many agents as assets and for almost any assignement of initial commodity endowments to agents, the resulting financial asset economy has $S - 1$ dimensions of real indeterminacy.

There is something of a surprise in this result. Indeed, we had initially conjectured that the number was $S - B$ (a number consistent also with Cass' example). As it turns out the dimension of indeterminacy is independent of the number of bonds B, as long as $0 < B < S$. If $B = 0$ the model is obviously determinate. If $B \geqslant S$, one can apply Arrow's [1] logic to show that all the equilibrium commodity allocations are Arrow-Debreu equilibrium allocations, and again there is no real indeterminacy. Theorem 1 points to an intriguing discontinuity. If markets are financially complete, then the model is determinate. Let just one financial asset be missing and the model becomes highly indeterminate. Thus, in this sense, the complete markets hypothesis ($B = S$) lacks robustness.[5] (Probably what this means is that the hypothesis has to be interpreted as $B > S$, i.e., one better have some redundancy.)

The idea behind the proof of Theorem 1 combines two essential ingredients. First, one can arbitrarily fix in advance the absolute level of commodity prices in terms of some numeraire independently across all the S states and still solve for equilibrium. The reason for this apparently puzzling phenomenon is that fixing the absolute price level in each state s is equivalent to transforming the financial assets into real assets that all deliver in the same numeraire commodity in each state s, and Geanakoplos-Polemarchakis [9] proved that numeraire asset economies always have equilibria. Second when $0 < B < S$, changes in the relative rates of inflation change the B-dimensional span of the assets. The set of B-dimensional subspaces of $\mathbb{R}^S$ has dimension $B(S - B)$. Hence if $0 < B < S$ there are enough distinct subspaces to be filled by the $S - 1$ degrees of freedom. Finally, we show that if there are more agents than assets, then a change of subspace typically means that the old equilibrium is no longer feasible.

There is no doubt that many contracts and financial securities in the world promise state contingent delivery in money, and not in real commodities. Moreover, there is little doubt that asset markets are incomplete. Nevertheless, it is perhaps worthwhile to make three brief comments about the robustness of our results. First, recall that we have taken the financial asset payoffs as fixed exogenously. We shall not make any further effort to explain how these payoffs are determined, or why others are missing. There are obvious reasons why some contracts are denoted in money terms, not the least of which is simplicity, and we see no reason why these monetary payoffs would change to fully offset any change in the expected absolute price level across the states. If they did, then they would indeed be

5. We emphasize that we are measuring degree of indeterminacy by number of dimensions. For other notions of size the story may well be different.

real assets. Or, at least, we see no reason why this would happen instantly to accommodate any unexpected shock.

Second, there is no doubt that there are a great number of real assets in the economy. One may conjecture that each independent real asset reduces by one the dimension of real indeterminacy in the economy, so that if there are A real assets, then there are only $S - 1 - A$ dimensions of real indeterminacy. But like our previous conjecture, this is incorrect. Theorem 2, in Section III, shows that as long as $A + B < S/2$, there are still $S - 1$ dimensions of real indeterminacy, independent of A or B. In summary: when markets are incomplete, the presence of financial assets creates an indeterminacy in competitive equilibrium allocations of a degree that does not depend on the absence of real assets.

Third, it is possible to give examples of financial asset payoff matrices R, that are not in general position and for which the dimension of real indeterminacy is less than $S - 1$. For example, if all the assets are Arrow securities, then there is typically no real indeterminacy. Different readers may have different views about which are the most salient financial asset payoff structures. We have accordingly, in Section III, introduced a simple formula that expresses the dimension of real indeterminacy typically associated with any financial asset payoff matrices R. We do not find that matrices R yielding no indeterminacy are more plausible than the R in general position (for which Theorem 1 yields maximal indeterminacy). In particular, our formula implies that as long as none of the rows of R is identically zero the dimension of real indeterminacy is always at least $S - B$ (see, also, Balasko-Cass [2] and Cass [5]).

II. The Model and Main Result

There are $L + 1$ physical commodities ($l = 0,\ldots, L$) and two dates. Spot trade tomorrow will take place under any of S states ($s = 1,\ldots, S$). Today there is trade on current goods and on B financial assets or bonds ($b = 1, \ldots, B$). Bonds pay money. We express their payoff by an $S \times B$ matrix with generic entry r_{sb}. We say that R is in *general position* if every submatrix of R has full rank.

There are $H + 1$ consumers ($h = 0, \ldots, H$). Every consumer h has a utility function u^h defined on $\mathbb{R}^{(L+1)(S+1)}_{++}$ and satisfying the standard differentiability, monotonicity, curvature, and boundary conditions needed to get a well-defined C^r differentiable excess demand function (see, for example, Mas-Colell [12, Chap. 2]). Note: The degree of differentiability r is assumed to be large enough for the subsequent transversality arguments to be justified. Every consumer also has an initial endowment vector $\omega^h \in \mathbb{R}_{++}^{(L+1)(S+1)}$. In this section, when we say that a property of economies $E = (R, u^h, \omega^h; h \in H)$ is *generic*, we mean that for any $\bar{R}$ and $\bar{u}^h$ there is an open, dense subset $\mathcal{D} \subset \mathbb{R}_{++}^{(L+1)(S+1)}$ whose complement has Lebesgue measure zero and such that the property applies to all economies $(\bar{R}, \bar{u}^h, \omega^h; h \in H)$ with endowment chosen in $\mathcal{D}$.

Definition 1. An allocation $(\bar{x}, \bar{y})$ of goods and bonds is a *financial asset equilibrium* for $E = (r, u^h, \omega^h; h \in H)$ if:

(i) $\Sigma_h \bar{x}^h = \Sigma_h \omega^h$, $\Sigma_h \bar{y}^h = 0$, and
(ii) there is a price system $p \in \mathbb{R}_{++}^{(L+1)(S+1)}$, $q \in \mathbb{R}^B$ such that for every h, $(\bar{x}^h, \bar{y}^h)$ maximizes $u^h(x^h)$ on

$$B^h(p,q) = \Big\{(x^h, y^h)\colon\ p_0 \cdot x_0^h + q \cdot y^h \leqslant p_0 \cdot \omega_0^h, \text{ and}$$

$$p_s \cdot x_s^h \leqslant p_s \cdot \omega_s^h + \sum_b y_b^h r_{sb}, \text{ all } s\Big\}.$$

Since the budget constraints imply that $S + 1$ of the market clearing conditions are redundant, there is in general some indeterminacy in the equilibrium allocations. If the indeterminacy affects only the holdings of bonds, y^h, then we call it *nominal indeterminacy.* Otherwise, we call it *real indeterminacy.* We are interested in the degree of real indeterminacy.

Theorem 1. If $0 < B < S$, R is in general position, and $H \geqslant B$, then, generically, there are $S - 1$ dimensions of real indeterminacy, i.e., the set of equilibrium allocations of commodities x contains the image of a C^1, one-to-one function with domain $\mathbb{R}^{S-1}$.

Proof of Theorem 1. The proof proceeds in four steps. The first introduces the notion of a real numeraire asset equilibrium and shows that the set of financial asset equilibrium can be parameterized as real numeraire asset equilibria, the parameter being an S-vector or prices of money in the different states. The second step gives sufficient conditions for the real numeraire asset equilibria corresponding to different parameters to be different. These conditions are of two types: (i) a full dimension requirement on the span of the vectors of individual demand for assets at equilibrium, and (ii) spanning requirements involving the return matrix and the particular parameter vector. Step 3 shows that the conditions of type (i) are generically satisfied if $H \geqslant B$. Step 4 shows that if R is in general position then the conditions of type (ii) are also satisfied for a $S - 1$ dimensional family of parameters. Combining we get the result.

Step 1. Given the prices p_{s0} of the zero commodity (or, equivalently, the price of money $\lambda_s = l/p_{s0}$ in terms of good 0) our system of financial assets is equivalent to a system of "real numeraire assets" where each asset pays in (equivalent worth) of the zero commodity. More precisely, given a matrix $\bar{R} = (\bar{r}_{sb})$, representing the payoffs of real assets in the numeraire (commodity zero) for each state s, let us define allocation $(\bar{x}, \bar{y})$ of goods and real assets to be a *real numeraire asset equilibrium* if (i) and (ii) of Definition 1 are satisfied, but with respect to the budget set

$$\bar{B}^h(p,q) = \Big\{(x, y)\colon\, p_0 \cdot x_0 + q \cdot y \leqslant p_0 \cdot \omega_0^h \text{ and}$$

$$p_s \cdot x_s \leqslant p_s \cdot \omega_s^h + p_{s0} \cdot \sum_b y_b^h \bar{r}_{sb} \text{ for all } s\Big\}.$$

It is easy to see that $(\bar{x}, \bar{y})$ is a financial asset equilibrium, with asset return R, if and only if $(\bar{x}, \bar{y})$ is a real numeraire asset equilibrium with $p_{s0} = 1$ for all $s \in S$ and asset return matrix $\bar{R} = \Lambda R$, where Λ is some diagonal $S \times S$ matrix having $\lambda_s > 0$ for all s.

Step 2. For any $S \times B$ matrix A, let us denote by $sp[A]$ the linear subspace of $\mathbb{R}^s$ spanned by the B columns of A.

Lemma 1. Let (x, y) and $(\hat{x}, \hat{y})$ be real numeraire asset equilibria for, respectively, $E = (\Lambda R, u^h, \omega^h; h \in H)$ and $\hat{E} = (\hat{\Lambda} R, u^h, \omega^h; h \in H)$. Suppose that dim $sp[y^1, \ldots, y^H] =$ dim $sp[\hat{y}^1, \ldots, \hat{y}^H] = B$ *and* $sp[\Lambda R] \neq sp[\hat{\Lambda} R]$. *Then* $x \neq \hat{x}$.

Proof. Consider the vectors $\{\Lambda R y^h\colon h \in H\}$ and $\{\hat{\Lambda} R \hat{y}^h\colon h \in H\}$. By hypothesis there is some h such that $\hat{\Lambda} R \hat{y}^h \neq \Lambda R y^h$. Suppose that $x^h = \hat{x}^h$. From the smoothness and boundary conditions on u^h, we must have that the goods equilibrium prices p and $\hat{p}$ are equal. But by Walras Law, which holds state by state, that implies $\hat{\Lambda} R \hat{y}^h = \Lambda R y^h$. Contradiction. ■

Step 3. Let M be the set of diagonal positive matrices. We will now establish a fairly intuitive fact. Namely that if $H \geqslant B$ then generically at equiblirium the vectors of individual assets demands span $\mathbb{R}^B$. More precisely, we show that generically there is an open nonempty subset $V \subset M$ and a C^1 parameterization of allocations $x(\Lambda)$, $y(\Lambda)$, $\Lambda \in V$, such that, first,

$(x(\Lambda), y(\Lambda))$ is a real numeraire asset equilibrium with return matrix ΛR, and, second, $y(\Lambda)$ satisfies the full dimension condition of Lemma 1 (and, therefore, if $\Lambda, \Lambda' \in V$ and $sp[\Lambda R] \neq sp[\Lambda' R]$ then $x(\Lambda) \neq x(\Lambda')$).

The proof uses standard transversality techniques. We will not repeat here the most familiar arguments.

Let $f(p, q, \Lambda, \omega)$ be the excess demand function from $P = \mathbb{R}^{L(S+1)}_{++} \times \mathbb{R}^B \times \mathbb{R}^S_{++} \times \mathbb{R}^{(L+1)(S+1)(H+1)}_{++}$ to $\mathbb{R}^{L(S+1)} \times \mathbb{R}^B$. Of course this function is not defined for all $q \in \mathbb{R}^B$ but only for those asset prices which satisfy a "non-arbitrage" condition.

Lemma 2. *f is a C^r function on the (nonempty) interior of its domain of definition. Moreover, $f(p, q, \Lambda, \omega) = 0$ if and only if p, q are real numeraire asset equilibrium prices for $E = (\Lambda R, u^h, \omega^h; h \in H)$. Also, $f(p, q, \Lambda, \omega) = 0$ implies that* rank $\partial_{\omega 0} f(p, q, \Lambda, \omega) = L(S + 1) + B$.

Proof. See Geanakoplos and Polemarchakis [9]. ■

Define now $g: P \times J \to \mathbb{R}^{L(S+1)} \times \mathbb{R}^B \times \mathbb{R}^B$, where J is the $B - 1$ sphere, by

$$g(p, q, \Lambda, \omega, z) = \left(f(p, q, \Lambda, \omega), \sum_{h=1}^{B} z_h y_1^h, \ldots, \sum_{h=1}^{B} z_h y_B^h \right),$$

where y_b^h is the hth consumer demand for bond b at (p, q, Λ, ω).

Lemma 3. *If* $g(p, q, \Lambda, \omega, z) = 0$ *then* rank $\partial_\omega g(p, q, \Lambda, \omega, z) = L(S + 1) + 2B$.

Proof. Let $(p, q, \Lambda, \omega, \bar{z}) \in g^{-1}(0)$. Because $\bar{z} \in J$ we know that $\bar{z}_h \neq 0$ for some h. Given Lemma 2 it suffices to show that for any $b = 1, \ldots, B$ there is some perturbation Δ^h and Δ^0 of the endowments of consumers h and 0 that leaves f and $y_{h', b'}$ unaffected for all $(h', b') \neq (h, b)$ but does change $y_{h, b}$. Let Δ^h be given by a decrease in ω^h_{00} of q_b and an increase in ω^h_{s0} of $\lambda_s R_{sb}$ for all $s = 1, \ldots, S$. Let Δ^0 be given by an increase in ω^h_{00} of q_b and a decrease in ω^h_{s0} of $\lambda_s R_{sb}$. Then consumer h decreases his demand $y_{h,b}$ by one unit and aggregate f is unaffected. ■

By the Transversality Theorem (see, e.g., Mas-Colell [12, Subsection 1.1]) for a.e. ω the sets $f_\omega^{-1}(0)$ and $g_\omega^{-1}(0)$ are C^r- manifolds of respective dimensions S and $S - 1$. By Sard's theorem (see reference above) the projection of $f_\omega^{-1}(0)$ on M has a regular value $\hat{\Lambda}$. From Geanakoplos and Polemarchakis [9] we know that $(\hat{p}, \hat{q}, \hat{\Lambda}, \omega) \in f^{-1}(0)$ for some $\hat{p}, \hat{q}$. Hence the regular value $\hat{\Lambda}$ is actually in the range of the projection. Therefore, from the Implicit Function Theorem, there are open sets $P'_\omega \subset P_\omega = \{(p, q, \Lambda): (p, q, \Lambda, \omega) \in P\}$, $V' \subset M$, and a C^1 function $\xi: V' \to P'_\omega$ such that $(p, q, \Lambda, \omega) \in f^{-1}(0) \cap (P_\omega \times \{\omega\})$ if and only if $\xi(\Lambda) = (p, q, \Lambda, \omega)$. Let $\bar{P}'_\omega \subset P_\omega$ be the closure of P'_ω. Then the projection of $g_\omega^{-1}(0) \cap (\bar{P}'_\omega \times J)$ on M is compact and so we can find a nonempty open set $V \subset V'$ which is disjoint from this projection. But this means that if $\Lambda \in V$ then the $\left\{y^h\right\}_{h=1}^{B}$ corresponding to $\xi(\Lambda)$ satisfy the spanning condition of Lemma 1. We have thus obtained the desired parameterization of equilibria.

Step 4. We now complete the proof by exploiting the hupotheses not yet used, namely that $B < S$ and R is in general position. We will see that this implies that $sp[\Lambda R] \neq sp[\Lambda' R]$ unless $\Lambda = \alpha\Lambda'$ for some $\alpha > 0$. Therefore, using Lemmata 1–3, the subset of M where $\lambda_I = 1$ provides our $S - 1$ parameterization. We begin by a preliminary lemma. We say that a collection of subspaces $L_1, \ldots, L_K \subset \mathbb{R}^B$ is linearly independent if $\Sigma_k y_k = 0$, $y_k \in L_k$ implies $y_k = 0$ for all h.

Lemma 4. Let R be an $S \times B$ matrix with nonzero rows and Λ a diagonal invertible matrix. If $sp[\Lambda R] \subset sp[R]$ then there are linearly independent subspaces $L_1, \ldots, L_K \subset \mathbb{R}^B$, such that, first, every row of R is contained in some subspace and, second, two rows belong to the same subspace if and only if the corresponding entries of Λ are equal.

Proof. The hypothesis $sp[\Lambda R] \subset sp[R]$is equivalent to the following: there is a $B \times B$ matrix Y such that $\Lambda R = RY$. This means that every row of R is an eigenvector of Y with the corresponding element of Λ being the eigenvalue. Given a linear transformation Y to each of its distinct real eigenvalues $\lambda_1, \ldots, \lambda_K$ we can associate the linear subspace $L_1, \ldots, L_K \subset \mathbb{R}^B$ where each L_k is spanned by the eigenvectors corresponding to λ_k. The collection $\{L_k\}$ is linearly independent (see, e.g., Halmos [10, p. 113]). ■

In our case we should have $K = 1$. Otherwise, because $B < S$ we would have a subspace $L \subset \mathbb{R}^B$ with dim $L < B$ but containing a number of rows of R larger than dim L. This contradicts the general position of R.

Summarizing, in our case $sp[\Lambda R] \subset sp[R]$ implies $\Lambda = \alpha I$, $\alpha > 0$. Let now $sp[\Lambda R] = sp[\Lambda' R]$. Then $sp[\Lambda'^{-1} \Lambda R] = sp[R]$ Hence $\Lambda = \alpha\Lambda'$ for some $\alpha > 0$, as we wanted to prove.

Remark 1. Observe that Theorem 1 and its corollary hold for any smooth utilities and asset matrix R, but only when there are more agents than assets, and only for a generic choice of endowments. There are good reasons for this, aside from the technical requirements of the transversality theorem. For example, if the endowment assignement were Pareto optimal, then there would be a unique (no trade) equilibrium allocation, no matter what the asset structure. Of course, if there is only one agent, then all endowment assignments are Pareto optimal, and Theorem 1 could not possibly apply. But if there is more than one agent, then generically an endowment assignement is not Pareto optimal.

Remark 2. Suppose that $B < H + 1 \leqslant S - 1$. In that case there are at least as many dimensions of indeterminacy as there are individual types. One would expect very often to find Pareto comparable financial equilibria.

Remark 3. If the assets were Arrow securities (i.e., every asset pays one dollar in a state of nature and nothing otherwise) then the model is generically determinate (see Geanakoplos and Polemarchakis [9]). Theorem 1 does not apply because R fails to be in general position when $B < S$. See Section III for more on this.

Remark 4. The conclusion of the theorem implies that the set of equilibrium real allocations x contains a nonempty $S - 1$ topological (i.e., C^0) manifold. The conclusion can be strengthened to C^1 manifold (one only needs to show that the derivative of the parameterization has full rank everywhere). Because nothing of economic substance is involved we skip the extra technical work.

Remark 5. The conclusion of the theorem can be sharpened when $H \geqslant SB$. In this case the entire set of equilibrium real allocations can be expressed as the differentiable one-to-one image of an $S - 1$ C^1 manifold (the observation parallel to Remark 4 also applies here). For the proof one considers the function $g: P \times J \to \mathbb{R}^{L(S+1)} \times \mathbb{R}^B \times \mathbb{R}^{SB}$, where J is the $S(B - 1)$ sphere, defined by

$$g(p, q, \Lambda, \omega, z^1, \ldots, z^S)$$

$$= \Big(f(p, q, \Lambda, \omega), \sum_{h=1}^{B} z_h^1 y_1^h, \ldots,$$

$$\sum_{h=1}^{B} z_h^1 y_B^h, \ldots, \sum_{h=1}^{B} z_h^S y_1^{(S-1)B+h}, \ldots, \sum_{h=1}^{B} z_h^S y_B^{(S-1)B+h} \Big)$$

Exactly as in the proof of Theorem 1, one shows that 0 is a regular value of g, hence for a generic ω, 0 is a regular value of g_ω. But this is impossible unless $g_\omega^{-1}(0) = \emptyset$ because the range of g_ω has greater dimension than its domain. If ω is generic for f and g we have then that $f_\omega^{-1}(0)$ is an S manifold. This yields that $E = \{(p, q, \Lambda, \omega) \in f_\omega^{-1}(0): \lambda_1 = 1\}$ is an $S - 1$ manifold. It is easily seen (use Lemma 1) that the real allocations corresponding to *any* two points in E are necessarily distinct.

Remark 6. At the risk of repeating ourselves (see the Introduction), it may be useful to devote a few words to understanding the failure of determinacy in Theorem 1 in the light of the conventional theory of regular economies (see, e.g., Mas-Colell [12, Sects. 8.2 and 8.3]). Formally, our general framework falls within the scope of the theory because money can be viewed as a physical commodity as any other and, similarly, its price is just one more relative price. The reason that the conclusions of the theory (i.e., generically the economy is determinate) do not apply is that technically our universe of admissible economies is degenerate. As long as consumers do not derive direct utility from money and the total endowment of the latter is kept equal to zero the excess demand for money remains null. In fact, the decisive factor is that the total endowment of money be zero (even if money is directly valued at equilibrium its consumption must be zero, i.e., consumers must be at the boundary of their consumption sets and, therefore, their demands for money may be locally insensitive to its relative price). If money aggregate endowments become positive the model is determinate (with unvalued money its price can only be zero).

But even if the general theory does not apply one could ask: How can the presence of an unvalued money commodity available in zero aggregate amount affect the equilibrium prices and allocations of the remaining commodities (we, after all, would not care if the indeterminacy fell entirely on the relative price of money)? The answer should be obvious: the relative price of the money commodity may have real (wealth) effects if consumers arrive to the market with nonzero entitlements of money (aggregating to zero). This can happen even in a one period two commodity world. What the incomplete markets contribute to the story is an endogenous reason (trade at time 0) for the nonzero individual endowments of money in the markets of period one.

Remark 7. Financial assets in our model yield payoffs in what might be called "inside money." The aggregate endowment of each asset, and the aggregate payoff in each state, is zero. This is, of course, of central importance to the indeterminacy that we find in financial assets markets since in any equilibrium for a finite horizon model outside money cannot be positively priced. However, in an infinite horizon model, like the overlapping generations model, it is possible to have nontrivial outside money. One could easily introduce uncertainty and financial assets that have nonzero aggregate supply into an overlapping generations economy. Indeed, what is called money in that model is the archetypical financial asset.

Remark 8. We have considered an economy with only two time periods. This is more general than it may appear at first sight. We could imagine an economy with many time periods, as in Debreu [6], where time and uncertainty resolve themselves as in a tree. If all assets are traded once and for all at data zero then the tree model can be regarded as a special case of our two period model with as many states of nature, as there are nodes in the tree (less one for date 0). The number of states of nature, hence the degree of indeterminacy, can grow geometrically with the length of the time horizon.

Remark 9. The possibility of combining Remarks 7 and 8 is intriguing. One is irresistibly lead to conjecture that in an overlapping generations economy with repeated moves of nature and incomplete *financial* markets there will be an infinity of dimensions of indeterminacy!

Remark 10. Although our theorem only holds for a generic set of endowments, one can guess that there are economies where across states the endowments and von Neumann-Morgenstern utilities are identical, and yet if markets are incomplete, the presence of financial assets creates $S - 1$ dimensions of real indeterminacy, i.e., $S - 1$ dimension of "sunspot" equilibria.

III. Refinements

In this section we present two refinements of Theorem 1. In the first (done in collaboration with J. Moore) we derive the general formula for the degree of indeterminacy when no restrictions whatsoever are imposed on the return matrix. In the second we discuss our problem when there are both financial and real assets.

For any $S \times B$ matrix R let $K(R)$ be the maximal number of linearly independent subspaces of $\mathbb{R}^B$ which satisfy the property that every subspace contains some nonzero row of R and that every row is contained in some subspace. Let $Z(R)$ be the number of rows which are identically zero. Then Theorem 1 can be strengthtened to:

Theorem 1′. If $H \geqslant B > 0$ then for any asset matrix R there are, generically, $S - K(R) - Z(R)$ dimensions of real indeterminacy.

Proof. We give the proof for the case $Z(R) = 0$. Accounting for the more general case is a trivial matter.

Denote $K = K(R)$. By hypothesis we can assume that the rows of R have been renumbered so that

$$R = \begin{bmatrix} R_1 \\ \vdots \\ R_K \end{bmatrix},$$

where the rows of the matrices $\{R_1, \ldots, R_K)$ span linearly independent subspaces of $\mathbb{R}^B$.

Let M_R be the collection of positive diagonal matrices having the properties that for every k the diagonal entries of the k-block are all equal (and denoted λ_k).

Lemma 5. We have $sp[\Lambda R] \subset sp[R]$ if and only if $\Lambda \in M_R$.

Proof. (i) Necessity. By Lemma 4 if two rows of some R_k are associated with different diagonal entries of Λ then it is possible to split the rows of R_k so that they generate two linearly independent subspaces. But this contradicts the maximality property of K. Hence $\Lambda \in M_R$.

(ii) Sufficiency. Let $\Lambda \in M_R$. Then $qR = 0 \Leftrightarrow q_k R_k = 0$ for all $k \Leftrightarrow \lambda_k q_k R_k = 0$ for all $k \Leftrightarrow q\Lambda R = 0$. The first and last implication follows from the linear independence of the subspaces generated by the rows of the different R_k (or $\lambda_k R_k$). Hence $sp[\Lambda R] = sp[R]$. ■

Because dim $M = S$ and dim $M_R = K$ Lemma 5 implies that there are precisely $S - K$ directions of perturbations Λ of the identity for which $sp[\Lambda R] \neq sp[R]$, Hence, noting that Steps 1 to 3 of the proof of Theorem 1 never use the hypothesis "$B < S$ and R is in general position," we have proved the more general Theorem 1′. ■

When R is in general position, Theorem 1 sharply distinguishes between the complete asset markets case ($B \geqslant S$), and the incomplete asset markets case ($B < S$) for which there are $S - 1$ dimensions of real indeterminacy. The general picture, however, is given by Theorem 1′.

Notice first that if the asset markets are complete, i.e., if all the rows of R are linearly independent, then $Z(R) = 0$ and $K(R) = S$, because the one-ditnensional subspaces spanned by each row separately are linearly independent. Hence the dimension of indeterminacy is $S - S - 0 = 0$, as it should be. Furthermore, if R consists of B distinct Arrow securities, then $K(R) = B$, $Z(R) = S - B$, and again the dimension of indeterminacy is $S - (S - B) - B = 0$, as it should be. If R is in general position and $0 < B < S$, then $K(R) = 1$ and $Z(R) = 0$, as we argued in the last two paragraphs of the proof of Theorem 1.

There is another special case which is of interest, and which indicates that the dimension of indeterminacy in practice is probably considerably less than $S - 1$. Suppose that S can be partitioned into disjoint subsets $S = S_1 \cup \cdots \cup S_K$ and that for each $i \leqslant K$, there is an asset in R which pays out one dollar if $s \in S_i$ and nothing otherwise. Then clearly $K(R) \geqslant K$, and so the dimension of indeterminacy is at most $S - K$. Observe, however, that the formula can also be made to yield a lower bound. As long as every state can be reached by at least one asset (i.e., $Z(R) = 0$) the dimension of indeterminacy must be at least $S - B \leqslant S - K(R)$.

Our second refinement takes as starting point the fact that in actuality there are both nominal and real assets. It may seem reasonable to conjecture that the larger is the proportion of real assets, the smaller is the indeterminacy associated with financial assets. However, we now show that the dimension of real indeterminacy is robust to the introduction of real securities, as long as markets are sufficiently incomplete. We will not make here an effort to get the best possible result.

In order to avoid the difficulties with existence that are known to plague models with real assets which yield vector-valued payoffs (see Hart [11]), we shall confine our attention to real numeraire assets, i.e., real assets that, for each state $s \in S$, pay only in commodity 0.

Let R be the $S \times B$ matrix representing the B financial assets and let $\overline{R}$ be the $S \times A$ matrix representing the real numeraire assets. Thus r_{sb} is the number of dollars paid by financial asset b in state s, and $\overline{r}_{sa}$ is the amount of good 0 paid by real asset a in state s.

The definition of an equilibrium is now a triple $(x, y, \overline{y})$ satisfying (i) and (ii) of Definition 1 with the budget set $B^h(p, q, \overline{q})$ defined as

$$\Big\{(x, y, \overline{y}): p_0 \cdot x_0 + q \cdot y + \overline{q} \cdot \overline{y} \leqslant p_0 \cdot \omega_0^h \quad \text{and}$$

$$p_s \cdot x_s \leqslant p_s \cdot \omega_s^h + \sum_{b=1}^{B} y_b^h r_{sb} + p_{s0} \sum_{a=1}^{A} y_a^h \overline{r}_{sa} \text{ for all } s\Big\}.$$

Theorem 2. Suppose that $B \geqslant 2$, $S > 2(A + B)$, and $H > A + B$. Then for a generic choice of matrices R and $\overline{R}$, there is a generic set of endowments such that each of the corresponding economies has $S - 1$ dimensions of real indeterminacy (in the sense of Theorem 1).

Proof. By following the logic of the proof of Theorem 1, it suffices to show that for a generic choice of matrices R and $\overline{R}$, there is no diagonal matrix $\Lambda \neq \sigma I$ with $\Lambda R \subset sp[R, \overline{R}]$

Let $\mathscr{D}$ be the set of $S \times (B + A)$ matrices $W = (R, \overline{R})$ which have rank $B + A$, satisfy $r_{sb} \neq 0$ for $b \in \{1, 2\}$and $s \in S$, and have $r_{s1}/r_{s'1} \neq r_{s2}/r_{s'2}$ for all $s \neq s' \in S$. Clearly, $\mathscr{D}$ is an open, dense subset of all $S \times (B + A)$ matrices. It has a complement of null measure. We shall show that there is a generic subset $\mathscr{D}' \subset \mathscr{D}$ of matrices $(R, \overline{R})$ for which only diagonal matrices Λ that are multiples of the identity satisfy $\Lambda R \subset sp[R, \overline{R}]$.

Suppose in particular that $\Lambda R^1 \in sp[R, \overline{R}]$ and $\Lambda R^2 \in sp[R, \overline{R}]$, where R^1, R^2 are, respectively, the first and second column of R. Since we can rewrite

$$\begin{bmatrix} \lambda_1 & & 0 \\ & \ddots & \\ 0 & & \lambda_S \end{bmatrix} \begin{bmatrix} r_{11} \\ \vdots \\ r_{S1} \end{bmatrix} = \begin{bmatrix} 1/r_{11} & & 0 \\ & \ddots & \\ 0 & & 1/r_{S1} \end{bmatrix}^{-1} \begin{bmatrix} \lambda_1 \\ \vdots \\ \lambda_S \end{bmatrix}$$

we must have two $(A + B)$-dimensional vectors z and $\hat{z}$ with

$$\begin{bmatrix} 1/r_{11} & & 0 \\ & \ddots & \\ 0 & & 1/r_{s1} \end{bmatrix} \big[R, \overline{R}\big] z = \begin{bmatrix} 1/r_{12} & & 0 \\ & \ddots & \\ 0 & & 1/r_{S2} \end{bmatrix} \big[R, \overline{R}\big] \hat{z}.$$

We know, of course, that $z = (\lambda, 0, \ldots, 0)$, $\hat{z} = (0, \lambda, 0, \ldots, 0)$ is always a solution for any λ. We show that for a generic choice of R and $\overline{R}$, there is no other choice of z and $\hat{z}$ that constitutes a solution.

Note first that if $S > 2(A + B)$, then there are more equations to satisfy than there are unknowns. It suffices to show, therefore, that eliminating from the domain the previous special configuration of z and $\hat{z}$, the above system of equations has zero as a regular value. That is, it suffices that given any R, $\bar{R}$ and solution $z \neq (\lambda, 0, \ldots, 0)$, $\hat{z} \neq (0, \lambda, 0, \ldots, 0)$, we can perturb any equation s by changing the R, $\bar{R}$, z, $\hat{z}$ in such a manner that the remaining equations are not disturbed. A routine application of the Transversality Theorem would then finish the proof.

Suppose that for a solution R, $\bar{R}$, z, $\hat{z}$, there is some k, $3 \leqslant k \leqslant A + B$ with either $z_k \neq 0$ or $\hat{z}_k \neq 0$ (or both). It follows that $z_k/r_{s1} \neq \hat{z}_k/r_{s2}$ for at least $S - 1$ of the S states. For any such state s, a small perturbation of w_{sk} (if $k \leqslant B$, $w_{sk} = r_{sk}$; if $B < k \leqslant A + B$, $w_{sk} = \bar{r}_{s,A+B-k}$) will change the sth equality without disturbing the rest. For the remaining state s_0, one can change z_1. That will affect every equality, including s_0; but this is clearly a perturbation with an effect which is independent of the other $S - 1$ perturbations.

Suppose alternatively that $z_k = \hat{z}_k = 0$ for all $k \geqslant 3$. Then we are back to exactly the framework of Step 4 in the proof of Theorem 1, with only two financial assets, and we know that there are no solutions to the system of equations except for $z = (\lambda, 0, \ldots, 0)$, $\hat{z} = (0, \lambda, \ldots, 0)$, which we have excluded from the domain.

Remark 11. More generally, we could, and should, also consider mixed assets which pay both in real commodities and in money. Once again there will be natural sufficient conditions guaranteeing that the dimension of indeterminacy is $S - 1$. For example, suppose that for each asset the states can be divided into those in which the asset pays in units of account and those in which the asset pays in numeraire commodities. Loans with collateral are of this type: there is a specified financial payment and a real collateral payoff in case of default (which here should be thought of as an exogenous event). One could also think of form-issued debt in similar terms. Let A be the total number of mixed assets. Suppose that for every $s \in S$ there are two assets and a collection $F(s) \subset S$ of at least $2A + 1$ states (including s) on which the two assets both pay in money. Then the proof of Theorem 2 does easily yield that there are $S - 1$ dimensions of real indeterminacy.

References

1. K. J. Arrow, Le rôle des valeurs boursières pour la répartition la meilleure des risques, *in* "Econometrie," pp. 41–47, Colloques Internationaux du CNRS, Vol. 11, CNRS, Paris, 1953.
2. Y. Balasko and D. Cass, "The Structure of Financial Equilibrium. I. Exogenous Yields and Unrestricted Participation," CARESS Working Paper No. 19186.
3. D. Cass, "Sunspots and Incomplete Financial Markets: The Leading Example," CARESS Working paper No. 84-06R, 1984, forthcoming in "Joan Robinson and Modern Economics" (G. Feiwel, Ed.).
4. D. Cass, "Competitive Equilibrium with Incomplete Financial Markets," CARESS Working Paper No. 84-09, 1984.
5. D. Cass, "On the 'Number' of Equilibrium Allocations with Incomplete Financial Markets," CARESS Discussion Paper No. 85-16, May 1985.
6. G. Debreu, "Theory of Value," Wiley, New York, 1959.
7. D. Duffie, Stochastic equilibria with incomplete financial markets, *J. Econ. Theory* 41 (1987), 405–416.
8. I. Fisher, The debt deflation theory of Great Depressions, *Econometrica* (1933), 337–357.
9. J. Geanakoplos and H. Polemarchakis, Existence, regularity, and constrained suboptimality of competitive portfolio allocations when asset markets are incomplete, *in* "Equilibrium Analysis," Essays in Honor of K. Arrow (W. Heller, R. Starr, and D. Starret, Eds.), Vol. III, Chap. 3, Cambridge Univ. Press, Cambridge, 1986.
10. P. Halmos, "Finite Dimensional Vector Spaces," Van Nostrand, London, 1958.
11. O. D. Hart, On the optimality of equilibrium when the market structure is incomplete, *J. Econ. Theory* 11 (1975), 418–443.
12. A. Mas-Colell, "The Theory of General Economic Equilibrium, A Differentiable Approach," Cambridge Univ. Press, Cambridge, 1985.
13. R. Radner, Existence of equilibrium of plans, prices, and price expectations, *Econometrica* 40 (1972), 289–303.
14. J. Werner, Equilibrium in economies with incomplete financial markets, *J. Econ. Theory* 36 (1985), 110–119.

— 7 —

Potential, Value, and Consistency[1]

SERGIU HART
ANDREU MAS-COLELL

Let P be a real-valued function defined on the space of cooperative games with transferable utility, satisfying the following condition: In every game, the marginal contributions of all players (according to P) are efficient (i.e., add up to the worth of the grand coalition). It is proved that there exists just one such function P—called *the potential*—and moreover that the resulting payoff vector coincides with the Shapley *value*. The potential approach is also shown to yield other characterizations for the Shapley value, in particular, in terms of a new internal *consistency* property. Further results deal with weighted Shapley values (which emerge from the above consistency) and with the nontransferable utility case (where the egalitarian solutions and the Harsanyi value are obtained).

Keywords: Shapley value, potential, consistency, n-person games, weighted values.

1. Introduction

Consider the problem of allocating some resource (or costs, profits, etc.) among n participants (economic agents, projects, departments, . . .). Assume that the situation is well described as an n-person game in character-

Econometrica 57 (1989), 589–614.

1. This paper supersedes our earlier papers, "Value and Potential: Marginal Pricing and Cost Sharing Reconciled" and "The Potential: A New Approach to the Value in Multi-Person Allocation Problems" (HIER, Harvard University, DP-1127, January, 1985 and DP-1157, June, 1985). We want to acknowledge useful discussions with Michael Maschler and Lloyd S. Shapley. Financial support by the National Science Foundation and by the U.S.-Israel Binational Science Foundation is gratefully acknowledged.

istic function form. The problem we address here is that of developing general principles for performing this allocation.

An approach with a long tradition in economics[2] would proceed by assigning to every player his direct marginal contribution to the grand coalition (i.e., the set of all players). It is obvious, however, that it is not possible in general to solve the problem in this way. This is simply because these marginal contributions may not add up to the worth of the grand coalition (namely, they will either be not feasible, or, if feasible, not efficient); from now on, we will refer to this "adding up" requirement simply as "efficiency."

In this paper we introduce a new analytical concept with clear affinities to the marginal contribution approach. We propose that to every allocation problem described by an n-person game be associated a single number (called the potential of the game) and that each player receive his marginal contribution (computed according to these numbers). The surprising fact is: the requirement that a feasible and efficient allocation (one that exactly shares everything available) should always be obtained determines the procedure uniquely. Namely, there exists just one such allocation procedure. Moreover, the resulting solution is well known: it is the Shapley (1953b) value.

In summary, our central result is the following:

Theorem A: There exists a unique real function on games—called the potential—such that the marginal contributions of all players (according to this function) are always efficient. Moreover, the resulting payoff vector is precisely the Shapley value.

This theorem is discussed at length in Section 2.

Although the potential is in its essence just a technical tool, it is nonetheless a powerful and suggestive one. In particular, the potential approach has suggested to us two further, substantive ways to characterize the Shapley value. The first (in Section 3) uses a "preservation of differences" principle, which is a straightforward generalization of the "divide the surplus equally" idea for two-person situations. The second (in Section 4) considers an internal "consistency" property: eliminating some of the players, after paying them according to the solution, does not change the outcome for any of the remaining ones. The main result here is as follows:

2. For a modern point of view, see, for example, Ostroy (1984).

Theorem B: Consider the class of solutions that, for two-person games, divide the surplus equally. Then the Shapley value is the unique consistent solution in this class.

In Section 4 we also discuss alternative consistency requirements (including one for various bargaining solutions).

Next, we broaden the class of consistent solutions by dropping symmetry. We obtain the following theorem (in Section 5):

Theorem C: Consider the class of solutions that, for two-person games, are efficient, invariant (in a very weak sense), and monotone. Then the weighted Shapley values are the only consistent solutions in this class.

The weighted Shapley values were introduced by Shapley (1953a) as nonsymmetric generalizations of the Shapley value. They are based on relative weights among the players. By our result, only one requirement—consistency—suffices to endogenously generate these weights (with the appropriate very weak initial conditions for two-person games being assumed).

Finally, we address also the nontransferable utility case (in Section 6): the potential approach leads naturally to the egalitarian solutions studied by Myerson (1980) and Kalai-Samet (1985). These can be seen as the first step in the construction of the Harsanyi (1963) NTU-value (Theorem D provides the exact characterization). Theorem E generalizes Theorem C. It shows that consistency (together with the correct initial conditions for two-person transferable utility games) characterizes the egalitarian solutions for nontransferable utility games too.

2. The Potential

A (cooperative) *game* (with side payments) is a pair (N, v), where N is a finite set of *players* and $v: 2^N \to R$ is[3] a *characteristic function* satisfying $v(\phi) = 0$. We will refer to a subset S of N as a *coalition*, and to $v(S)$ as the *worth* of S. Given a game (N, v) and a coalition S, we write (S, v) for the *subgame* obtained by restricting v to subsets of S only (i.e., to 2^S).

Let $\mathbf{G}$ be the set of all games. Given a function $P: \mathbf{G} \to R$ which associates

3. R denotes the real line.

a real number[4] $P(N, v)$ to every game (N, v), the *marginal contribution* of a player in a game is defined to be

$$D^iP(N, v) = P(N, v) - P(N\backslash\{i\}, v),$$

where (N, v) is a game and $i \in N$; recall that the (sub)game $(N\backslash\{i\}, v)$ is the restriction of (N, v) to $N\backslash\{i\}$.

A function P: $\mathbf{G} \rightarrow R$ with $P(\phi, v) = 0$ is called a *potential function* if it satisfies the following condition:

$$\sum_{i \in N} D^i P(N, v) = v(N) \tag{2.1}$$

for all games (N, v). Thus, a potential function is such that the allocation of marginal contributions (according to the potential function) always adds up exactly to the worth of the grand coalition. From now on, we refer to this property as "efficiency."

Theorem A: There exists a unique potential function P. For every game (N, v), the resulting payoff vector $(D^iP(N, v))_{i \in N}$ coincides with the Shapley value of the game. Moreover, the potential of any game (N, v) is uniquely determined by (2.1) *applied only to the game and its subgames (i.e., to (S, v) for all $S \subset N$).*

Let $Sh^i(N, v)$ denote the *Shapley value* of player i in the game (N, v). We thus have

$$D^iP(N, v) = Sh^i(N, v)$$

for every game (N, v) and each player i in N.

Proof of Theorem A: Formula (2.1) can be rewritten as

$$P(N, v) = \frac{1}{|N|}\left[v(N) + \sum_{i \in N} P(N \setminus \{i\}, v)\right]. \tag{2.2}$$

Starting with $P(\phi, v) = 0$, it determines $P(N, v)$ recursively. This proves the existence of the potential function P, and moreover that $P(N, v)$ is uniquely determined by (2.1) (or (2.2)) applied to (S, v) for all $S \subset N$.

Next, express (N, v) as a linear combination of unanimity games: $v = \Sigma_{T \subset N} a_T u_T$, where u_T is the *T-unanimity* game, defined by $u_T(S) = 1$ if S contains T and $= 0$ otherwise (it is well known that this decomposition

4. We write $P(N, v)$ rather than the more cumbersome $P((N, v))$.

exists and is unique). Define $d_T = a_T/|T|$ (this is Harsanyi's "dividend" to the members of coalition T), and put

$$P(N, v) = \sum_T d_T. \tag{2.3}$$

It is easily checked that (2.1) is satisfied by this P; hence (2.3) defines the unique potential function. The result now follows since $Sh^i(N, v) = \Sigma_{\{T:\, i \in T\}} d_T$ (i.e., T ranges over all subsets T of N that contain i).[5] *Q.E.D.*

For further interpretation of the potential we derive an explicit formula ((2.1) defines it only implicitly). Consider the following (standard) model of choosing a random subset S of a given set N with $n = |N|$ elements: First, a size $s = 1, 2, \ldots, n$ is chosen randomly (with probability $1/n$ each). Second, a subset S of N of size s is chosen randomly (with probability $1/\binom{n}{s}$), where $s = |S|$). Equivalently, one can order the n elements (there are $n!$ orders), choose a cutting point s (there are n choices), and take the first s elements in the order. Let E denote expectation with respect to this probability distribution.

Proposition 2.4: Let P be the potential function. Then

$$P(N, v) = E\left[\frac{|N|}{|S|} v(S)\right]$$

for every game (N, v).

Proof: Using the explicit probabilities described above, we have to show that

$$P(N, v) = \sum_{S \subset N} \frac{(s-1)!(n-s)!}{n!} v(S), \tag{2.5}$$

where $n = |N|$ and $s = |S|$. The marginal contributions D^iP of (2.5) are easily seen to coincide with the Shapley value; therefore (2.5) is the potential function. *Q.E.D.*

Thus, the potential is the *expected normalized worth;* equivalently, the per-capita potential $P(N, v)/|N|$ equals the average per-capita worth

5. An alternative proof is to show that $(D^i P)_i$ satisfy the standard axioms for the Shapley value (this is done inductively, using (2.1) or (2.2); see Hart and Mas-Colell (1988), which includes alternative proofs for other results here, as well as further discussions).

$v(S)/|S|$. Hence, the potential provides a most natural one-number summary of the game. The Shapley value is known to be an expected marginal contribution. We obtain it here as a marginal contribution to an expectation (the potential).[6]

Remark 2.5: Our approach can be regarded as a *new characterization* for the Shapley value. Only one axiom, (2.1), suffices. Moreover, only the game itself and its subgames have to be considered; this is important particularly in applications, where typically only *one* specific problem is considered. In contrast, the standard axiomatizations require the application of the axioms to a large class of games (e.g., all games; or, all simple games; etc.) in order to uniquely determine it for any single game. Finally, note that although marginal contributions appear implicitly in the formulae for the Shapley value, our approach uses this principle explicitly (i.e., in the axioms), and in a very simple form.

Remark 2.6: Formula (2.2) yields a simple and straightforward recursive procedure for the computation of the potential, and thus, a fortiori, for the Shapley value of the game as well as of all its subgames. It seems to be a most efficient algorithm for computing Shapley values.

Remark 2.7: It is clear that P is, formally, just a mathematical potential function for the Shapley value (taking discrete rather than infinitesimal differences); this explains our choice of name for it. Thus, the Shapley value vector function is the (discrete) gradient of the potential function.[7] Moreover, if we do not require $P(\phi, v) = 0$, then P is only determined up to an additive constant (which of course does not change the payoff vectors).[8]

6. The idea of reversing the order of integration and differentiation has been fruitfully used by Mertens (1980) in the context of values of nonatomic games.

7. This "summability" condition that is satisfied by the Shapley value can be stated as follows: For every game (N, v) and every $S \subset N$ with $|S| = s$

$$\sum_{t=1}^{S} Sh^{i_t}(\{i_1, i_2, \ldots, i_t\}, v)$$

is the same for all orderings $i_1, i_2, \ldots, i_s$ of the elements of S (this number is just $P(S,v)$). This condition may be viewed as "path-independence."

8. We note that D'Aspremont, Jacquemin, and Mertens (1984) have recently defined a class of real functions on games: "aggregate (power) indices." It can be easily checked that the potential function does not belong to this class. However, it would be included if their "Normalization axiom" is dropped (the potential function is obtained when, in their notations, $d\mu(\alpha) = d\alpha/\alpha$—i.e., ν is the Lebesgue measure).

Remark 2.8: One may regard P as an operator that associates to each game (N, v) another game (N, Pv), given by $(Pv)(S) = P(S, v)$ for all $S \subset N$. It is easily checked that this is a linear, positive, and symmetric operator; it is one-to-one and onto, and its fixed points are exactly the inessential (additive) games. Additional properties are implied by these; for example, if (N, v) is a market game (i.e., totally balanced), then $Pv \leqslant v$.

The potential approach can be extended to games with a continuum of players. We consider here only the finite type case: the game is given by a pair (f, z), where $f: R^n_+ \to R$ (with $f(0) = 0$) is the characteristic function, and $z = (z^1, \ldots, z^n) \in R^n_+$ represents the grand coalition (z^i being the "number" of players of type i). This is, for instance, the standard setup for cost allocation problems, f being either the "cost" or the "production" function (see Billera and Heath (1982) and Mirman and Tauman (1982)). A (type-symmetric) feasible and efficient solution to (f, z) is then a vector $\phi(f, z) \in R^n$ with the property $z \cdot \phi(f, z) = f(z)$. A *potential* for f is a differentiable function $F: R^n_+ \to R$ such that for any z we have $z \cdot \partial F(z) = f(z)$, where $\partial F(z)$ is the gradient of F at z. All the results in this section generalize: Given f, there is one and only one potential[9] F. Moreover, the solution $\phi(f, z) = \partial F(z)$ corresponds to the Aumann-Shapley (1974) value for games with a continuum of players (it is thus called "the Aumann-Shapley price vector"). Hence, this is the only integrable vector field ψ_f on R^n_+ satisfying $z \cdot \psi_f(z) = f(z)$ for all $z \in R^n_+$. Finally, the potential F can be derived explicitly and takes a familiar diagonal form (compare with Proposition 2.4):

$$F(z) = \int_0^1 \frac{1}{t} f(tz)\, dt.$$

3. Preservation of Differences

We now take a different tack to the payoff allocation problem. It will, however, lead us to the same solution.

Consider the grand coalition N in the game (N, v). Suppose we are given constants d^{ij} for all $i, j \in N$ which are *compatible,* in the sense that $d^{ii} = 0$,

9. Minor regularity conditions are required; for example, f continuous and $f(z) = 0(||z||)$ as $z \to 0$ suffice. (See the formula for F below.)

$d^{ij} = -d^{ji}$ and $d^{ij} + d^{jk} = d^{ik}$ for all $i, j, k \in N$. We say that a payoff vector $(x^i)_{i \in N}$ *preserves differences according to* $\{d^{ij}\}$ if

$$x^i - x^j = d^{ij} \quad \text{for all } i, j.$$

It is trivial to verify that, given compatible constants $\{d^{ij}\}$, there exists a single *efficient* payoff vector x that preserves differences:

$$x^i = \frac{1}{|N|}\left[v(N) + \sum_j d^{ij}\right], \quad x^j = x^i - d^{ij}.$$

To have a well defined solution we therefore only need to specify, for every game, differences $\{d^{ij}\}$. We do this recursively. Suppose that payoffs have been determined for all strict subgames of (N, v); let $x^i(S)$ be the payoff of player i in (S, v), for $i \in S \subset N$. Then we convene that the difference d^{ij} to be preserved is:

(3.1) $$d^{ij} = x^i(N\backslash\{j\}) - x^j(N\backslash\{i\});$$

that is, the difference between what i would get if j was not around and what j would get if i was not around. (It will be seen below that these differences are indeed compatible.) This seems to be a most natural way to compare the "relative position" (or, "relative strengths") of the players. It is notable that, as it will be shown below, one obtains *a unique efficient outcome which simultaneously preserves all these differences.*[10]

The preservation of differences principle can be regarded as a straightforward generalization of the "equal division of the surplus" idea for two-person problems. Indeed, in that case, the payoffs are

$$x^i \equiv x^i(\{i, j\}) = v(\{i\}) + \frac{1}{2}[v(\{i, j\}) - v(\{i\}) - v(\{j\})],$$

thus

$$x^i - x^j = v(\{i)) - v(\{j\}) = x^i(\{i\}) - x^J(\{j\}),$$

or: differences are preserved.

Observe that the proposed solution satisfies

(3.2) $$\sum_{i \in N} x^i(N) = v(N), \quad \text{and}$$

(3.3) $$x^i(N) - x^i(N\backslash\{j\}) = x^j(N) - x^J(N\backslash\{i\}).$$

10. One wants to preserve differences rather than, say, ratios, since the resulting outcome should not depend on the choice of origin of a player's utility scale.

Condition (3.3) has already been used by Myerson (1980), under the name of "balanced contributions." The mathematically inclined reader will recognize it as a finite difference analog of the Frobenius integrability condition (i.e., symmetry of the cross partial derivatives), which suggests that the solution admits a potential function. Conversely, if a solution is generated by a potential, then (3.3) is clearly satisfied.

Theorem 3.4: The construction of the above solution, according to the principle of preservation of differences, is well-defined. Moreover, the solution is generated by a potential function—thus it coincides with the Shapley value.

Proof: Consider an n-person game (N, v). Assume, by induction, that the solution has been already determined for all strict subgames of (N, v), and that moreover

$$x^i(S) = P(S) - P(S \backslash \{i\}) \tag{3.5}$$

for all $i \in S \subset N$, $S \neq N$ (where P is the unique potential function, and $P(S)$ is short for $P(S, v)$). Note that (3.5) is of course satisfied for singletons ($|S| = 1$) by efficiency (3.2).

Applying (3.5) to coalitions of size $n - 1$ implies that the differences d^{ij} are indeed compatible, since

$$\begin{aligned} d^{ij} &= x^i(N\backslash\{j\}) - x^j(N\backslash\{i\}) \\ &= [P(N\backslash\{j\}) - P(N\backslash\{i, j\})] - [P(N\backslash\{i\}) - P(N\backslash\{i, j\})] \\ &= P(N\backslash\{j\}) - P(N\backslash\{i\}). \end{aligned}$$

Therefore $(x^i(N))_i$ is well defined, and we have

$$x^i(N) - x^j(N) = d^{ij} = P(N\backslash\{j\}) - P(N\backslash\{i\}).$$

Fix i, and average over j:

$$x^i(N) - \frac{1}{n} v(N) = \frac{1}{n} \sum_{j \in N} P(N\backslash\{j\} - P(N\backslash\{i\})$$

(by (3.2)). Thus

$$x^i(N) + P(N\backslash\{i\}) = \frac{1}{n}\left[v(N) + \sum_{j \in N} P(N\backslash\{j\})\right].$$

By (2.2), the right-hand side is exactly $P(N)$; thus (3.5) is satisfied for $S = N$ also. *Q.E.D.*

4. Consistency

This section is devoted to a characterization of the Shapley value by means of an (internal) consistency property. Such an approach may be traced back to Harsanyi (1959). It has been successfully applied to a wide variety of solution concepts: Davis and Maschler (1965), Sobolev (1975), Lensberg (1982), Balinsky and Young (1982), Thomson (1984), Peleg (1985, 1986), Aumann and Maschler (1985), Moulin (1985), etc. We will discuss the connections between some of these approaches later in this section.

The consistency requirement may be described informally as follows: Let ϕ be a function that associates a payoff to every player in every game. For any group of players in a game, one defines a reduced game among them by giving the rest of the players payoffs according to ϕ. Then ϕ is said to be consistent if, when it is applied to any reduced game, it yields the same payoffs as in the original game. The various consistency requirements differ in the precise definition of the reduced game (i.e., exactly how are the players outside being paid off).

Formally, let ϕ be a function defined on **G**, the set of all games (see Section 2), with $\phi(N, v)$ a vector in R^N for all (N, v) in **G**; such a function is called a *solution function.* We will write $\phi^i(N, v)$ for its i th coordinate; thus, $\phi(N, v) = (\phi^i(N, v))_{i \in N}$.

Let ϕ be a solution function, (N, v) a game, and $T \subset N$. We define the *reduced game* (T, v_T^ϕ) as follows:

$$(4.1) \qquad v_T^\phi(S) = v(S \cup T^c) - \sum_{i \in T^c} \phi^i(S \cup T^c, v), \quad \text{for all} \quad S \subset T,$$

where $T^c = N \backslash T$. A solution function ϕ is *consistent* if, for every game (N, v) and every coalition $T \subset N$, one has

$$(4.2) \qquad \phi^j(T, v_T^\phi) = \phi^j(N, v), \quad \text{for all} \quad j \in T.$$

The interpretation is as follows. Given the solution function ϕ, a game (N, v) and a coalition $T \subset N$, the members of T (or, more precisely, every subcoalition of T) need to consider the total payoff remaining after paying the members of T^c according to ϕ. To compute the worth of a coalition $S \subset T$ (in the reduced game), we assume that the members of $T \backslash S$ are not present; in other words, one considers the game $(S \cup T^c, v)$, in which payoffs are distributed according to ϕ. The appropriateness of this definition of reduced games depends, of course, on the particular situation being modelled; more specifically, on the concrete assumptions underlying the determination of the characteristic function. An example will be discussed later on in this section.

Note that one usually deals with efficient solution functions. In this case, (4.1) can be rewritten as

$$(4.3) \qquad v_T^\phi(S) = \sum_{i \in S} \phi^i(S \cup T^c, v).$$

Furthermore, if ϕ is an efficient and consistent solution function, then necessarily

$$v_T^\phi(T) = \sum_{j \in T} \phi^j(T, v_T^\phi) = \sum_{j \in T} \phi^j(N, v) = v(N) - \sum_{i \in T^c} \phi^i(N, v),$$

which is exactly (4.1) for $S = T$. If one wants the definition of $v_T^\phi(S)$ to be always according to the same rule, then (4.1) results for all $S \subset T$.

The following lemma shows that whether ϕ is consistent or not may be determined by considering only singleton coalitions T^c.

Lemma 4.4: ϕ *is a consistent solution function if and only if (4.2) is satisfied for all games* (N, v) *and all* $T \subset N$ *with* $|T^c| = 1$.

Proof: Use induction on the size of T^c. *Q.E.D.*

The first result is as follows:

Proposition 4.5: *The Shapley value is a consistent solution function.*

Proof: We will use the potential P, as defined by (2.1). Let (N, v) be a game and let $i \in N$; we will write v_{-i} for $v_{N\backslash\{1\}}^\phi$, where $\phi = Sh$. Since Sh is efficient, we have for all $S \subset N\backslash\{i\}$:

$$\begin{aligned} v_{-i}(S) &= v(S \cup \{i\}) - Sh^i(S \cup \{i\}, v) = \sum_{j \in S} Sh^j(S \cup \{i\}, v) \\ &= \sum_{j \in S} [P(S \cup \{i\}, v) - P(S \cup \{i\}\backslash\{j\}, v)]. \end{aligned}$$

By Theorem A, formula (2.1) applied to $(N\backslash\{i\}, v_{-i})$ and all its subgames *uniquely* determines their potential. Comparing this with the above equalities, we obtain that $P(S, v_{-i})$ and $P(S \cup \{i\}, v)$ may differ only by a constant[11]

$$P(S, v_{-i}) = P(S \cup \{i\}, v) + c.$$

11. From $P(\phi, v_{-i}) = 0$ it follows that $c = -P(\{i\}, v)$; this is however not needed in the sequel.

Thus

$$Sh^j(N\backslash\{i\}, v_{-i}) = P(N\backslash\{i\}, v_{-i}) - P(N\backslash\{i, j\}, v_{-i})$$
$$= P(N, v) - P(N\backslash\{j\}, v)$$
$$= Sh^j(N, v). \qquad \text{Q.E.D.}$$

Next, we will show that the property of consistency is essentially equivalent to the existence of a potential function; thus, consistency almost characterizes the Shapley value. "Almost" refers to the "initial conditions," namely, the behavior of the solution for two-person games.[12]

A solution function ϕ is *standard for two-person games* if

$$(4.6) \qquad \phi^i(\{i, j\}, v) = v(\{i\}) + \tfrac{1}{2}\,[v(\{i, j\}) - v(\{i\}) - v(\{j\})]$$

for all $i \neq j$ and all v. Thus, the "surplus" $[v(\{i, j\}) - v(\{i\}) - v(\{j\})]$ is equally divided among the two players. Most solutions satisfy this requirement, in particular, the Shapley value and the nucleolus.

Theorem B: Let *ϕ be a solution function. Then:*
(i) *ϕ is consistent; and*
(ii) *ϕ is standard for two-person games;*
if and only if ϕ is the Shapley value.

Proof:[13] One direction is immediate (recall Proposition 4.5).

For the other direction, assume ϕ satisfies (i) and (ii). We claim first that ϕ is efficient, i.e.,

$$(4.7) \qquad \sum_{i \in N} \phi^i(N, v) = v(N)$$

for all (N, v). This indeed holds for $|N| = 2$ by (4.6). Let $n \geqslant 3$, and assume (4.7) holds for all games with less than n players. For a game (N, v) with $|N| = n$, let $i \in N$; by consistency

$$\sum_{j \in N} \phi^j(N, v) = \sum_{j \in N\backslash\{i\}} \phi^j(N\backslash\{i\}, v_{-i}) + \phi^i(N, v),$$

12. In contrast to the potential approach (in Section 2), consistency requires one to consider a large domain of games rather than a single one. Moreover, the reduced games may not share some of the properties of the original game; e.g., super-additivity (consider the three-person majority game).

13. An independent proof (not based on the potential) has been communicated to us by Michael Maschler. For yet another proof of uniqueness, see Lemma 6.8.

where $v_{-i} \equiv v^{\phi}_{N\setminus\{i\}}$. By assumption, ϕ is efficient for games with $n - 1$ players; thus

$$= v_{-i}(N\setminus\{i\}) + \phi^i(N, v) = v(N)$$

(by definition of v_{-i}). Therefore ϕ is efficient for all $n \geqslant 2$.

Finally, for $n = 1$, we have to show that $\phi^i(\{i\}, v) = v(\{i\})$. Indeed, let $v(\{i\}) = c$, and consider the game $(\{i, j\}, \overline{v})$ (for some $j \neq i$), with $\overline{v}(\{i\}) = \overline{v}(\{i, j\}) = c$, $\overline{v}(\{j\}) = 0$. By (ii), $\phi^i(\{i, j\}, \overline{v}) = c$ and $\phi^j(\{i, j\}, \overline{v}) = 0$; hence $\overline{v}_{-j}(\{i\}) = c - 0 = c = v(\{i\})$, and $c = \phi^i(\{i, j\}, \overline{v}) = \phi^i(\{i\}, \overline{v}_{-j}) = \phi^i(\{i\}, v)$ by consistency. This concludes the proof of the efficiency of ϕ.

Next, we show that ϕ admits a potential. To that end, define a real function Q on the set of all games of at most two players by

$$\begin{aligned} Q(\phi, v) &= 0, \\ Q(\{i\}, v) &= v(\{i\}), \\ Q(\{i, j\}\ v) &= \tfrac{1}{2}[v(\{i\}) + v(\{j\}) + v(\{i, j\})], \end{aligned}$$

for all v and all $i \neq j$. It is straightforward to check that, for all (N, v) with $|N| = 1, 2$:

$$\phi^i(N, v) = Q(N, v) - Q(N\setminus\{i], v) \tag{4.8}$$

for all $i \in N$.

We will now show that Q can be extended to all games (N, v), in such a way that (4.8) always holds. Together with efficiency (4.7), this implies (2.1); therefore Q is actually the potential P, and ϕ is the Shapley value.

We again use induction: Let $n \geqslant 3$, and assume Q has been defined, and moreover satisfies (4.8), for all games of at most $n - 1$ players. Fix a game (N, v) with $|N| = n$. We have to show that

$$\phi^i(N, v) + Q(N\setminus\{i\}, v)$$

is the same for all $i \in N$ (and this will then be $Q(N, v)$). Let $i, j \in N$, $i \neq j$, and let $k \in N$, $k \neq i, j$ (such a k exists since $|N| \geqslant 3$).

We have (by consistency and (4.8) for $n - 1$)

$$\begin{aligned} &\phi^i(N, v) - \phi^j(N, v) \\ &\quad = \phi^i(N\setminus\{k\}, v_{-k}) - \phi^j(N\setminus\{k\}, v_{-k}) \\ &\quad = [Q(N\setminus\{k\}, v_{-k}) - Q(N\setminus\{i, k\}, v_{-k})] \\ &\qquad - [Q(N\setminus\{k\}, v_{-k}) - Q(N\setminus\{j, k\}, v_{-k})] \\ &\quad = [Q(N\setminus\{j, k\}, v_{-k}) - Q(N\setminus\{i, j, k\}, v_{-k})] \\ &\qquad - [Q(N\setminus\{i, k\}, v_{-k}) - Q(N\setminus\{i, j, k\}, v_{-k})]. \end{aligned}$$

Apply again (4.8) (for $n - 2$) and consistency:

$$\begin{aligned} &= \phi^i(N\backslash\{j, k\}, v_{-k}) - \phi^j(N\backslash\{i, k\}, v_{-k}) \\ &= \phi^i(N\backslash\{j\}, v) - \phi^j(N\backslash\{i\}, v) \\ &= [Q(N\backslash\{j\}, v) - Q(N\backslash\{i, j\}, v)] \\ &\quad - [Q(N\backslash\{i\}, v) - Q(N\backslash\{i, j\}, v)] \\ &= Q(N\backslash\{j\}, v) - Q(N\backslash\{i\}, v), \end{aligned}$$

where we used (4.8) once more (for $n - 1$). This completes the proof. *Q.E.D.*

Remark 4.9: Theorem B (and all our other results on consistency in this paper) applies to any fixed finite number of players; i.e., if one considers only games with at most n players, then consistency together with the appropriate initial condition suffice to characterize the solution. This should be contrasted with other consistency results, e.g., Sobolev (1975), Lensberg (1982), Peleg (1986), where an unbounded number of players is needed.

The standard solution for two-person games is very natural; it may, however, be derived from more basic postulates.

A solution function ϕ is *transferable-utility invariant* (TU-invariant)[14] if, for any two games (N, v) and (N, u) (with the same set of players) and real constants $a > 0$ and $\{b^i\}_{i \in N}$,

$$u(S) = av(S) + \sum_{i \in S} b^i \quad \text{for all} \quad S \subset N$$

implies

$$\phi^i(N, u) = a\phi^i(N, v) + b^i \quad \text{for all} \quad i \in N.$$

(Note: (N, v) and (N, u) are called *TU-equivalent* games.) TU-invariance requires, first, that a change in scale *common* to all players should affect the solution accordingly; and second, that adding a fixed amount, whenever a player i appears, should lead to just adding this amount to his final payoff.

A solution function ϕ satisfies the *equal treatment property* if, for any game (N, v) and any two players $i, j \in N$,

$$v(S \cup \{i\}) = v(S \cup \{j\}) \quad \text{for all} \quad S \subset N\backslash\{i, j\}$$

14. This is sometimes referred to as "strategic invariance;" we prefer to call it "TU-invariance" instead, in order to avoid confusion with "NTU-invariance," where scales of different players may change independently.

implies

$$\phi^i(N, v) = \phi^j(N, v).$$

Note that, for two-person games, this amounts to just

$$\phi^i(\{i, j\}, v) = \phi^j(\{i, j\}, v) = \tfrac{1}{2}v(\{i, j\})$$

whenever $v(\{i\}) = v(\{j\})$.

Theorem B′: *Let ϕ be a solution function. Then:*

(i) *ϕ is consistent;*

(ii) *for two-person games:* (a) *ϕ is efficient,* (b) *ϕ is TU-invariant,* (c) *ϕ satisfies the equal treatment property;*

if and only if ϕ is the Shapley value.

Proof: Efficiency (a) and equal treatment (c) imply that

$$\phi^i(\{i, j\}, cu_{\{i, j\}}) = \tfrac{1}{2}c$$

for all $i \neq j$ and all real c, where $u_{\{i, j\}}$ denotes the $\{i, j\}$-unanimity game. Using this for $c = 1, 0$ and -1 together with TU-invariance (b) yields (4.6), and Theorem B applies. *Q.E.D.*

It is instructive to compare our approach with Sobolev's (1975) consistency treatment of the (pre)nucleolus. The two concepts of consistency—namely, (4.2)—are the same. The difference lies in the definition of the reduced game.[15] Following Davis and Maschler (1965), Sobolev uses the following definition for the reduced game (i.e., instead of (4.1)):

$$(4.10)\qquad v_T^{\phi}(S) = \begin{cases} \sum_{j \in T} \phi^j(N, v) & S = T; \\ \operatorname*{Max}_{Q \subset T^c} \left\{ v(S \cup Q) - \sum_{i \in Q} \phi^i(N, v) \right\}, & S \neq T, \phi; \\ 0, & S = \phi. \end{cases}$$

There are two important distinctions between the two definitions of the reduced game: first, the maximum over all $Q \subset T^c$ in (4.10) vs. just $Q = T^c$

15. Also, Sobolev uses the other assumptions (e.g., symmetry) in an essential way (for all n).

in (4.1); and second, the payoffs to the players in T^c are taken, for every S, according to the solution of the game (N, v) in (4.10), and according to the solution of the subgame $(S \cup T^c, v)$ in (4.1).

These two differences may be understood as follows. The first indicates that, in (4.10), some sort of strategic freedom is available to each coalition S: they may choose which of the members of T^c (if any) to take along and pay them according to the solution. In comparison, in (4.1) it is assumed that *all* members of T^c have to be paid off. The second difference lies in the way these payoffs are computed. In (4.10), the members of T^c are paid according to the solution *allocation* (of the grand coalition), whereas in (4.1) according to the solution *function,* each time applied to the appropriate situation (namely, the solution allocation of the coalition $S \cup T^c$).

As we have already stated above, which definition is more appropriate will depend on the context being modeled (and the way the characteristic function is defined). An example where our definition seems natural is the problem of allocating joint costs among several "projects" (or, departments, tasks, etc.); these are now interpreted as the players. It is to be emphasized that these cost imputations are *not* meant to be "efficiency prices" (i.e., usable to make investment decisions on which of the projects to undertake); in fact, except in trivial cases, cost allocations satisfying the adding-up condition would generate inefficient decisions if used as investment guides. In summary the problem is: we are given a fixed set of projects, and required (by the legal and/or administrative environment—e.g., the tax authorities) to obtain an exact distribution of total costs. To this end, one should, of course, take into account all available information, in particular, the cost of any subset S of projects (assuming that these are the *only* ones to be undertaken). We assume that this information (that may not be always easy to get) is available to the "accountants" in charge of computing the cost imputations.

What does consistency mean in this framework? Consider a multi-state company and a restricted set T of projects, say those in Tennessee. For every subset $S \subset T$ of Tennessee's projects, the local accountant has to determine their cost, assuming that these are the only projects in Tennessee to be undertaken. In addition, there is the set T^c of all the projects outside Tennessee (which are not in the domain of the local "gedanken experiment"). The cost of S is therefore the amount imputed to S by the general accounting procedure under consideration (solution function) when the projects to be implemented are $S \cup T^c$. This is exactly formula (4.3), and consistency requires that the local accountant's imputation be no different than the one obtained (for the Tennessee projects) by the general (national) accountant.

As suggested by the comparison of (4.1) and (4.10) there are two additional possible definitions of the reduced game. One is

$$v_T^{\phi}(S) = v(S \cup T^c) - \sum_{i \in T^c} \phi^i(N, v) \tag{4.11}$$

for all $S \subset T$. This definition is similar to (4.1) according to the first criterion mentioned above (no maximum), and similar to (4.10) according to the second criterion ($\phi^i(N)$ rather than $\phi^i(S \cup T^c)$). The resulting consistency has been studied by Moulin (1985). He shows (Lemma 6 there) that it characterizes the so-called "equal allocation of nonseparable costs."

Finally, the fourth possible definition of a reduced game is:

$$v_T^{\phi}(S) = \begin{cases} v(N) - \sum_{i \in T^c} \phi^i(N, v) & S = T, \\ \operatorname*{Max}_{Q \subset T^c} \left\{ v(S \cup Q) - \sum_{i \in Q} \phi^i(S \cup Q, v) \right\}, & S \subset T. \end{cases} \tag{4.12}$$

However, there is no solution function that is standard for two-person games and is consistent according to (4.12).[16]

5. Weights

The result in the previous section, Theorem B′, characterizes the Shapley value by means of consistency[17] together with natural initial conditions for two-person games. In this section we drop the equal treatment property, and characterize the resulting consistent solutions. They turn out to be the *weighted Shapley values*, introduced by Shapley (1953a) (see also Owen (1968), Shapley (1981), Kalai and Samet (1987)).

To present these solutions, we will first assume that a collection of

16. A counterexample is provided by the following four-person game: $N = \{1, 2, 3, 4\}$; player 4 is a null player; $v(\{1, 2\}) = v(\{2, 3\}) = 12$, $v(\{1, 3\}) = 24$, $v(\{1, 2, 3\}) = 27$, and $v(S) = 0$ for all other $S \subset \{1, 2, 3\}$. One may, however, restrict the class of games under consideration. For example, a class of games can be found (containing, in particular, all convex games), at which the Shapley value is the unique solution that is standard for two-person games and consistent according to (4.12).

17. From now on we will only deal with the original consistency (i.e., according to (4.1)).

weights is *exogenously* given. All the results of the previous sections (with the appropriate modifications) will be seen to remain valid in this more general setup: they are all related to a notion of weighted potential. We will then obtain our main result (Theorem C): consistency and the initial conditions for two-person games of Theorem B′, with equal treatment replaced by monotonicity, imply the existence of weights such that the solution is precisely the corresponding weighted Shapley value. Thus, weights are obtained *endogenously*, from the solution itself.

Formally, suppose first that for each player i we are given a weight $w^i > 0$; let[18] $w = (w^i)_i$. These weights can be interpreted as "a-priori measures of importance;" they are taken to reflect considerations *not* captured by the characteristic function. For example, we may be dealing with a problem of cost allocation among investment projects. Then the weights w^i could be associated to the profitability of the different projects. In a problem of allocating travel costs among various institutions visited (cf. Shapley (1981)), the weights may be the number of days spent at each one.

In line with the above interpretation, we would now desire that in any unanimity game the worth be distributed among the players in proportion to their weights. In a unanimity game, every player has exactly the same (marginal) contribution. Therefore, to obtain the w-proportional allocation we need to weight the marginal contribution of each player i by his w^i.[19]

We are thus led to the definition of a *w-potential* P_w, as a function $P_w\colon G \to R$ with $P_w(\phi, v) = 0$, satisfying the following condition:

$$\sum_{i \in N} w^i D^i P_w(N, v) = v(N) \tag{5.1}$$

for all games (N, v). The potential of Section 2 of course corresponds to $w^i = 1$ for all i.

Theorem 5.2: For every collection $w = (w^i)_i$ of positive weights there exists a unique w-potential function P_w. Moreover, the resulting solution function, associating the payoff vector $(w^i D^i P_w(N, v))_{i \in N}$ to the game (N, v), coincides

18. One may think of a universe U of players, such that N is always a finite subset of U (cf. Shapley (1953b)).

19. A player with weight 2 may be thought of (in some contexts) as two players with weight 1 (see Shapley (1981)).

with the w-weighted Shapley value Sh_w. Finally, P_w can be computed recursively by the formula

$$P_w(N,v) = \left[v(N) + \sum_{i \in N} w^i P_w(N \backslash \{i\}, v)\right] \Big/ \sum_{i \in N} w^i.$$

Proof: The only thing to check is that the w-potential indeed yields the weighted Shapley value Sh_w. It is easily seen (by the recursive formula above, for example) that $P_w(N, v) = 0$ if (N, v) is the null game (with $v(S) = 0$ for all S), and $P_w(N, cu_N) = c/\Sigma_{i \in N} w^i$ for an N-unanimity game (c is a constant). Therefore $w^i D^i P_w(N, cu_N) = cw^i/\Sigma_{i \in N} w^i = Sh^i_w(N, cu_N)$. Together with additivity (shown inductively using the recursive formula), the proof is completed. *Q.E.D.*

Remark 5.3: We are assuming throughout that all the weights are positive numbers. One may extend this to allow zero or infinite weights (see Kalai and Samet (1987)).

Note that the w-potential is (positively) homogeneous of degree minus one on the weights, and the corresponding payoff vector (the weighted Shapley value) is homogeneous of degree zero. One could thus think of the w-potential as being measured in per-unit-weight terms. It is to be again emphasized that the weights are unrelated to the characteristic function and that the players (or projects, etc.) are to be regarded as indivisible.

An explicit formula for the w-potential function (of which Proposition 2.4 is a special case) is the following: For each i, let X^i be a random variable with distribution function $\text{Prob}(X^i \leqslant t) = t^{w^i}$ for all $t \in [0,1]$, and assume all the X^i's are independent. For every $t \in [0,1]$, let $N(t) = \{i \in N \mid X^i \leqslant t\}$ (if one interprets X^i as the arrival time of i, then $N(t)$ is the set of players that arrived up to time t; see Owen (1968)). Then it may be checked that

$$P_w(N,v) = E\left[\int_0^1 \frac{1}{t} v\big(N(t)\big)\, dt\right].$$

The preservation of differences principle (see Section 3) also applies here; it just becomes

$$(5.4) \quad \frac{1}{w^i} x^i(N) - \frac{1}{w^j} x^j(N) = d^{ij} = \frac{1}{w^i} x^i(N \backslash \{j\}) - \frac{1}{w^j} x^j(N \backslash \{i\})$$

(thus, the differences between the normalized—i.e., per-unit-weight—payoffs are preserved).

Up to this point the introduction of weights may appear as a rather ad-hoc construction. For practical purposes we may want to add some flexibility to the Shapley value, obtaining nonsymmetric generalizations. But, why do it in this particular way? We will now see that the consistency postulate leads us directly to this class of weighted Shapley values.

Recall the definition of a reduced game (4.1) and the consistency postulate (4.2). The first observation is the following:

Proposition 5.5: For every collection of positive weights $w = (w^i)_i$, *the corresponding weighted Shapley value* Sh_w *is a consistent solution function.*

Proof: Mutatis mutandis as for Proposition 4.5, replacing the potential by the w-potential. *Q.E.D.*

Next, we generalize the standard solution to two-person games ("divide the surplus equally") in a nonsymmetric way as follows:

Let $w = (w^i)_i$ be positive weights. Then a solution function ϕ is *w-proportional for two-person games* if

$$\phi^i(\{i, j\}, v) = v(\{i\}) + \frac{w^i}{w^i + w^j}[v(\{i, j\}) - v(\{i\}) - v(\{j\})] \tag{5.6}$$

for all $i \neq j$.

We then have the generalization of Theorem B.

Theorem 5.7: Let ϕ *be a solution function and* w *a collection of positive weights. Then:*

(i) ϕ *is consistent; and*

(ii) ϕ *is w-proportional for two-person games;*

if and only if ϕ *is the w-weighted Shapley value* Sh_w.

Proof: Mutatis mutandis as for Theorem B. *Q.E.D.*

In Theorem 5.7 the weights are introduced exogeneously through the initial conditions for two-person games. We shall now see that, with some very weak conditions on the behavior of the solution for two-person games, consistency actually enables us to rule out all but the proportional solutions. Thus, we obtain a completely endogenous generation of the weights, and, a fortiori, of the weighted Shapley values.

We need the following definition (which will actually be applied only to two-person games). A solution function ϕ is *monotonic* if, for any two games (N, v) and (N, u),

$$u(N) > v(N) \quad \text{and} \quad u(S) = v(S) \quad \text{for all} \quad S \neq N$$

imply

$$\phi^i(N, u) > \phi^i(N, v) \quad \text{for all} \quad i \in N.$$

Thus, if the (grand coalition's) total payoff increases, but nothing else changes, then every player should get an increase in his own payoff. Note that, for two-person games, monotonicity is just $\phi^i(\{i, j\}, u) > \phi^i(\{i, j\}, v)$ and $\phi^j(\{i, j\}, u) > \phi^j(\{i, j\}, v)$ when $u(\{i\}) = v(\{i\})$, $u(\{j\}) = v(\{j\})$ and $u(\{i, j\}) > v(\{i, j\})$.

We now obtain the main result of this section.

Theorem C: Let ϕ be a solution function. Then:

(i) *ϕ is consistent; and*

(ii) *for two-person games:* (*a*) *ϕ is efficient,* (*b*) *ϕ is TU-invariant,* (c) *ϕ is monotonic;*

if and only if there exist positive weights $w = (w^i)_i$ such that ϕ is the w-weighted Shapley value.

The proof of Theorem C can be found in the Appendix. It is important to point out that, in contrast to Theorem B′, the conditions (ii) (a)–(c) for two-person games do *not* imply that ϕ is a w-proportional solution for $n = 2$. One needs to use consistency repeatedly (for $n = 3$) in order to obtain the initial condition for $n = 2$.

6. Nontransferable Utility

We now extend the potential function approach to the general case, where utility need not be additively transferable.

A *nontransferable-utility game*—an *NTU-game,* for short—is a pair (N, V), with $V(S)$ a subset of R^S for all coalitions S of N. The interpretation is that $x = (x^i)_{i \in S} \in V(S)$ if and only if there is an outcome attainable by the coalition S, whose utility to member i of S *is* x^i. From now on, we will refer to games in **G** as *TU* or *transferable-utility games.* A TU-game (N, v) in **G** corresponds to the NTU-game (N, V), where

$$V(S) = \left\{x \in R^S: \sum_{i \in S} x^i \leqslant v(S)\right\}.$$

We make the following (standard) assumptions: All sets $V(S)$ are (i) non-empty, (ii) not the whole space (R^S), and (iii) comprehensive.

How are marginal contributions computed in an NTU-game? Clearly, the formula $P(N, V) - P(N\backslash\{i\}, V)$ leads to interpersonal comparison of utilities, since all players use the same number $P(N, V)$. It is thus appropriate to use it only when the weights of the players are equal.

This suggests the following construction: First, use the potential function approach to obtain, for each collection $w = (w^i)_i$ of positive weights, a solution x_w. Second, require that w represent the appropriate marginal rates of efficient substitution between the players, *at* x_w. This is a standard procedure for obtaining solutions in the nontransferable utility case. One first assumes that the utility scales of the players are comparable (according to the weights w) and then requires that these are indeed the "right" weights at the resulting solution. This makes the final solution correspond to a fixed-point (of the mapping $w \rightarrow x_w \rightarrow w'$), and, most important, independent of rescaling utilities (for each player separately).

Formally, let $w = (w^i)_i$ be a collection of positive weights. The *w-potential function* P_w associates with every NTU-game (N, V) a real number $P_w(N, V)$, such that

$$(6.1) \qquad (w^i D^i P_w(N, V))_{i \in N} \in \text{bd}\, V(N)$$

where bd $V(N)$ denotes the (Pareto efficient) boundary of the set $V(N)$ (recall assumption (iii)); without loss of generality, let again $P_w(\phi, V) = 0$. Thus, (6.1) is the exact counterpart in the NTU-case of (2.1) (or, (5.1)) in the TU-case: the vector of (rescaled) marginal contributions is efficient.[20]

Theorem 6.2: For every collection $w = (w^i)_i$ of positive weights there exists a unique w-potential function on the class of NTU-games. Moreover, the resulting solution function, associating the payoff vector $(w^i D^i P_w(N, v))_{i \in N}$ to the NTU-game (N, V), coincides with the w-egalitarian solution.

The *egalitarian solutions* have been studied by Myerson (1980) and Kalai and Samet (1985). They may be viewed as the first step in the construction of the *Harsanyi (NTU-) value* (cf. Harsanyi (1963)). Note that the w-egalitarian solution coincides with the w-Shapley value for TU-games.

Proof: Assumptions (i)–(iii) above imply that, for each S, the set $V(S)$ is

20. It is easy to check that, for TU-games (represented in NTU form as above), we obtain the same w-potential of Section 5.

bounded from above in any strictly positive direction, hence bd $V(S)$ intersects any such line in a unique point. The proof proceeds by induction as in Theorem A, yielding $P_w(N, V)$ as the unique t such that $y + tw \in$ bd $V(N)$, where $y^i = -w^i P_w(N\backslash\{i\}, V)$ for all $i \in N$. This construction is easily seen to give the w-egalitarian solution (see Kalai and Samet (1985)). *Q.E.D.*

A (Pareto) efficient payoff vector $x \in$ bd $V(N)$ is called *w-utilitarian* (for given positive weights $w = (w^i)_i$) if it maximizes the sum of the utilities over the feasible set $V(N)$, rescaled according to w:

$$\sum_{i \in N} \frac{1}{w^i} x^i \geqslant \sum_{i \in N} \frac{1}{w^i} y^i \quad \text{for all} \quad y \in V(N).$$

Finally, x is a *Harsanyi* (1963) *value* if there exist weights w such that it is simultaneously w-egalitarian and w-utilitarian. We thus finally obtain the following theorem.

Theorem D: For every NTU-game (N, V), the payoff vector $x \in R^N$ is a Harsanyi value[21] *of (N, V) if and only if there exist positive weights $w = (w^i)_i$ such that*

$$\frac{1}{w^i} x^i = D^i P_w(N, V), \quad \textit{for all } i \in N, \textit{ and}$$

$$\sum_{i \in N} \frac{1}{w^i} x^i \geqslant \sum_{i \in N} \frac{1}{w^i} y^i, \quad \textit{for all } y \in V(N).$$

Since the egalitarian solutions are obtained by the potential approach, one is naturally led to investigate whether they can in fact be also characterized by the preservation of differences principle, and by consistency. It should come as no surprise that these alternative approaches apply just as in the TU-case.

In what concerns the preservation of differences principle, for every given set of weights w, it is easily seen that it uniquely determines the w-egalitarian solution. Of course, the efficiency condition (3.2) becomes $x(N) \in \text{bd}V(N)$; as for (3.3) and its weighted counterpart (5.4), they remain unchanged. The application of this principle when the utilities are

21. We actually obtain only the nondegenerate Harsanyi values (i.e., those corresponding to positive weights). Again, this can be easily fixed (see Remark 5.3).

not transferable and are only assumed to be comparable (by the weights w) seems to yield a good foundation for the egalitarian solutions (and thus, a fortiori, for the Harsanyi value). In particular, there is no need for "dividends" (cf. Harsanyi (1963)) which accumulate (and which may sometimes be negative); instead, one considers *only the relative positions* of the players as more and more coalitions are taken into account.

Next, consider *consistency*. The definition of the reduced game is the natural extension of (4.1) to the NTU-case (see Maschler and Owen (1986)). For every solution function ϕ, a game (N, V), and a subset $T \subset N$, the reduced game (T, V_T^{ϕ}) is defined by:

$$V_T^{\phi}(S) = \left\{x \in R^S : \left(x, \left(\phi^i(S \cup T^c, V)\right)_{i \in T^c}\right) \in V(S \cup T^c)\right\}$$

for all $S \subset T$. Thus, $V_T^{\phi}(S)$ is the S-section of $V(S \cup T^c)$ when the coordinates of all players outside T are fixed at their solution imputation (in the $S \cup T^c$ subgame). The solution function ϕ is consistent if

$$\phi^j(T, V_T^{\phi}) = \phi^j(N, V)$$

for all games (N, V) and all $j \in T \subset N$. Note that, in the case of an NTU-game that corresponds to a TU-game, this definition coincides with the one used in Sections 4 and 5.

From now on we restrict our attention to *nonlevel* NTU-games, defined as those games (N, V) such that, for all $S \subset N$, if $x, y \in \text{bd}V(S)$ and $x \geqslant y$, then $x = y$. This condition has been widely used in the study of NTU-games; it means that strong and weak efficiency coincide. We use it here in order to guarantee that, if $x = (x^i)_{i \in N}$ is efficient, then $x_S = (x^i)_{i \in S}$ is efficient in the corresponding reduced game.

The main result we obtain is the following:

Theorem E: *Let ϕ be a solution function on nonlevel NTU-games and let w be a collection of positive weights. Then:*

(i) *ϕ is consistent; and*

(ii) *ϕ is the w-proportional solution for two-person TU-games;*

if and only if ϕ is the w-egalitarian solution.

Remark 6.3: The initial conditions (ii) apply to two-person *TU*-games only. It is therefore noteworthy that the consistency requirement enables one to obtain the w-egalitarian solution without having to assume it for two-person *NTU*-games, where it is a much stronger primitive postulate than in the two-person TU-case. Moreover, it will be seen in Lemma 6.9 below that, if one assumes that ϕ is the w-egalitarian solution for two-person *NTU*-games (i.e., two-person bargaining problems) then consistency easily implies that it coincides with the w-egalitarian solution for all games (this is exactly as in the TU-case: see Theorems B and 5.7).

Remark 6.4: Theorems B′ and C show that the initial conditions (ii) follow from more primitive postulates using consistency. In particular, the weights are obtained *endogenously.* For example, Theorems C and E imply the following theorem:

Theorem 6.5: Let ϕ be a solution function on nonlevel NTU-games. Then:

(i) *ϕ is consistent;*

(ii) *for two-person TU-games: (a) ϕ is efficient, (b) ϕ is TU-invariant, (c) ϕ is monotonic;*

if and only if there exist positive weights w such that ϕ is the w-egalitarian solution.

Remark 6.6: Since Theorem E and Theorem 6.5 characterize only the w-egalitarian solutions, it follows that one cannot require in addition invariance under *independent* utility rescalings (i.e., where players may rescale differently). This has been shown by Maschler and Owen (1986), who also discuss a weaker form of consistency.

Proof of Theorem E: It will follow from Lemmata 6.7–6.9 below.

Lemma 6.7: The w-egalitarian solution is consistent.

Proof: The proof follows the same lines as Proposition 4.5, since it is generated by a potential function. *Q.E.D.*

Lemma 6.8: There exists a unique consistent solution that is w-egalitarian on two-person NTU-games.

Proof: One may use an argument similar to that in Theorem B. We will instead provide an alternative direct proof (which also works for Theorem B, and Theorem 5.7). First, as in the TU-case, one shows by induction that if a solution function is efficient for $n = 2$ and is consistent, then it is efficient for all n.

Let ϕ and ψ be two solution functions satisfying the hypothesis, and assume by induction that they coincide for all games of at most $n - 1$ players. Let (N, V) be an n-player game, and let $i, j \in N$, $i \neq j$. Consider the two reduced games $(\{i, j\}, V^{\phi}_{\{i,j\}})$ and $(\{i, j\}, V^{\psi}_{\{i,j\}})$, which we will denote for short by V^{ϕ} and V^{ψ}. They coincide for singletons (by induction, since only $n-1$ players matter); therefore, since ϕ is w-egalitarian for two-person games, we have $\phi^i(V^{\phi}) \gtreqless \phi^i(V^{\psi})$ if and only if $\phi^j(V^{\phi}) \gtreqless \phi^j(V^{\psi})$ (both lie on the same strictly positive ray). Now $\phi = \psi$ for two-person games, and both ϕ and ψ are consistent; therefore

$$\phi^i(V) = \phi^i(V^{\phi}) \gtreqless \phi^i(V^{\psi}) = \psi^i\,(V^{\psi}) = \psi^i(V)$$

if and only if, similarly,

$$\phi^j(V) \gtreqless \psi^j(V).$$

This applies to any two players, and both $\phi(V)$ and $\psi(V)$ are efficient; thus $\phi(V) = \psi(V)$. Q.E.D.

Lemma 6.9: If ϕ is consistent and w-proportional for two-person TU-games, then it is w-egalitarian for two-person NTU-games.

Proof: Let $(\{i, j\}, V)$ be a two-person NTU-game. Denote $z = (z^i, z^j) = \phi(V)$. We will show that z is the w-egalitarian solution of V, i.e., $w^i(z^i - \alpha^i) = w^j(z^j - \alpha^j)$, where $\alpha^i = \sup\,\{x^i\colon x^i \in V(\{i\})\}$ and $\alpha^j = \sup\,\{x^j\colon x^j \in V(\{j\})\}$.

Define a three-person NTU-game $(\{i, j, k\}, U)$ as follows:

$$U(S) = \begin{cases} V(S), & \text{if } S \subset \{i, j\}, \\ \left\{ x \in R^S\colon \sum\limits_{i \in S} x^i \leqslant \sum\limits_{i \in S} \alpha^i \right\}, & \text{otherwise} \end{cases}$$

where we put $\alpha^k = 0$.

Let $(y^i, y^j, y^k) = \phi(U)$.

Consider the reduced games $U^{\phi}_{-i}, U^{\phi}_{-j}, U^{\phi}_{-k}$. It can be easily checked that they all correspond to TU-games, which we will denote by u_{-i}, u_{-j}, and u_{-k}, respectively:[22]

$$u_{-i}(j) = z^j,\ u_{-i}(k) = 0,\ u_{-i}(jk) = y^j + y^k,$$

$$u_{-j}(i) = z^i,\ u_{-j}(k) = 0,\ u_{-j}(ik) = y^i + y^k,$$

$$u_{-k}(i) = \alpha^i,\ u_{-k}(j) = \alpha^j,\ u_{-k}(ij) = y^i + y^j.$$

22. We write $u(i)$, $u(ij)$, etc., instead of $u(\{i\})$, $u(\{i, j\})$, ...

Therefore, since ϕ is consistent, and is the w-proportional solution for two-person TU-games, we obtain:

$$w^j(y^j - z^j) = w^k(y^k - 0),$$
$$w^i(y^i - z^i) = w^k(y^k - 0),$$
$$w^i(y^i - \alpha^i) = w^j(y^j - \alpha^j),$$

from which the required equality follows. *Q.E.D.*

School of Mathematical Sciences, Beverly and Raymond Sackler Faculty of Exact Sciences, Tel Aviv University, 69978 Tel-Aviv, Israel

and

Department of Economics, Harvard University, Cambridge, MA 02138, U.S.A.

Manuscript received July, 1985; revision received July, 1988.

Appendix

Proof of Theorem C: In view of Theorem 5.7, we have to prove that (i) and (ii) imply that ϕ is a proportional solution for two-person games: namely, there exist positive weights $\{w^i\}$, such that (5.6) holds.

Let $i \neq j$, and consider the $\{i, j\}$-unanimity game $(\{i, j\}, u_{\{i,j\}})$: $u_{\{i,j\}}(i) = u_{\{i,j\}}(j) = 0$ and $u_{\{i,j\}}(i, j) = 1$. Put

$$\alpha^i_{\{i,j\}} = \phi^i(\{i, j\}, u_{\{i,j\}}), \quad \text{and}$$

(A.1) $$\beta^i_{\{i,j\}} = -\phi^i(\{i, j\}, -u_{\{i,j\}}).$$

By efficiency (ii) (a), we have

(A.2) $$\alpha^i_{\{i,j\}} + \alpha^j_{\{i,j\}} = \beta^i_{\{i,j\}} + \beta^j_{\{i,j\}} = 1.$$

A two-person game $(\{i, j\}, v)$ is *essential* if the "surplus" $\sigma = v(i, j) - v(i) - v(j)$ is not zero; it is then strategically equivalent to either $u_{\{i,j\}}$ or $-u_{\{i,j\}}$, depending on the sign of σ. By TU-invariance (ii)(b) and (A.1) we therefore obtain, for any essential game $(\{i, j\}, v)$,

(A.3) $$\phi^i(\{i, j\}, v) = v(i) + \delta^i[v(i, j) \quad v(i) - v(j)],$$

where $\delta^i = \alpha^i_{\{i,j\}}$ if the surplus $\sigma > 0$ and $\delta^i = \beta^i_{\{i,j\}}$ if $\sigma < 0$.

By monotonicity (ii)(c), the coefficient δ^i of $v(i, j)$ must be positive; thus

(A.4) $$\alpha^i_{\{i,j\}} > 0, \qquad \beta^i_{\{i,j\}} > 0.$$

Let $(\{i, j\}, v)$ be a two-person inessential game: $v(i, j) = v(i) + v(j)$. Let $v_\varepsilon^+, v_\varepsilon^-$ be defined as follows ($\varepsilon > 0$):

$$v_\varepsilon^+ (i, j) = v(i, j) + \varepsilon,$$
$$v_\varepsilon^- (i, j) = v(i, j) - \varepsilon,$$
$$v_\varepsilon^+ (i) = v_\varepsilon^-(i) = v(i),$$
$$v_\varepsilon^+ (j) = v_\varepsilon^+(j) = v(j).$$

Monotonicity yields:

$$\phi^i(\{i, j\}), v) < \phi^i(\{i, j\}, v_\varepsilon^+) = v(i) + \varepsilon\alpha^i_{\{i,j\}}.$$
$$> \phi^i(\{i, j\}, v_\varepsilon^-) = v(i) + \varepsilon\beta^i_{\{i,j\}}.$$

As ε decreases to 0 we obtain $\phi^i(\{i, j\}, v) = v(i)$. Thus, formula (A.3) applies to inessential games as well.

In particular, we note that (A.3) and (A.4) imply

(A.5) $\phi^i(\{i, j\}, v) - v(i)$ and $\phi^j(\{i, j\}, v) - v(j)$ have always the same sign ($+$, $-$, *or* 0).

Lemma A.6: There exist positive numbers $\{w^i\}_i$ *such that*

$$\alpha^i_{\{i,j\}} = \frac{w^i}{(w^i + w^j)} \textit{ for all } i \neq j.$$

Proof: Fix a player k, and define:

$$w^k = 1,$$
$$w^i = \frac{\alpha^i_{\{i,k\}}}{\alpha^k_{\{i,k\}}} \text{ for } i \neq k.$$

We have to show that

(A.7) $$\frac{\alpha^i_{\{i,j\}}}{\alpha^j_{\{i,j\}}} = \frac{w^i}{w^j} = \frac{\alpha^i_{\{i,k\}}}{\alpha^k_{\{i,k\}}} \cdot \frac{\alpha^k_{\{j,k\}}}{\alpha^j_{\{j,k\}}};$$

together with (A.2), it will imply our result.

Let (N, v) be the $\{i, j, k\}$-unanimity game: $N = \{i, j, k\}$, $v(N) = 1$, $v(S) = 0$ for all $S \neq N$. Let $x = \phi(N, v)$ be its solution.

Consider now the reduced game $(N\backslash\{i\}, v_{-i})$, *where* $v_{-i} \equiv v^{\phi}_{N\backslash\{i\}}$:

$$v_{-i}(j, k) = v(i, j, k) - \phi^i(\{i, j, k\}, v) = v(N) - x^i = x^j + x^k,$$

$$v_{-i}(j) = v(i, j) - \phi^i(\{i, j\}, v) = 0 - 0 = 0$$

(use (A.3), or note directly that $(\{i, j\}, v)$ is the null game, thus inessential). Similarly,

$$v_{-i}(k) = 0.$$

Consistency and (A.3) now imply that

$$x^j = \phi^j(N, v) = \phi^j(N\backslash\{i\}, v_{-i}) = \delta^j(x^j + x^k), \text{ and}$$

(A.8) $$x^k = \phi^k(N, v) = \phi^k(N\backslash\{i\}, v_{-i}) = \delta^k(x^j + x^k),$$

where $\delta^j = \alpha^j_{\{j,k\}}$ or $\beta^j_{\{j,k\}}$, and $\delta^k = \alpha^k_{\{j,k\}}$ or $\beta^k_{\{j,k\}}$. Therefore x^j and x^k have the same sign ($+$, $-$, or 0): recall (A.5) or (A.4).

This holds for any two players in N; therefore x^i, x^j, and x^k all have the same sign. Together with $x^i + x^j + x^k = v(N) = 1$, this yields $x^i, x^j, x^k > 0$. From (A.8) we obtain

$$\alpha^j_{\{j,k\}} = \delta^j = \frac{x^j}{x^j + x^k} \quad \text{and}$$

$$\alpha^k_{\{j,k\}} = \delta^k = \frac{x^k}{x^j + x^k}.$$

This is true for any two players in N, from which (A.7) follows immediately. *Q.E.D.*

Lemma A.9: There exist positive numbers $\{u^i\}_i$ *such that*

$$\beta^i_{\{i,j\}} = \frac{u^i}{u^i + u^j} \quad \textit{for all } i \neq j.$$

Proof: The same as for Lemma A.6, using $-v$ instead of v. *Q.E.D.*

Lemma A.10: There exists a positive number ρ *such that* $u^i = \rho w^i$ *for all i.*

Proof: Let i, j, k be all different, and consider the following game: $(\{i, j, k\}, v)$, with $v(i, k) = w^i + w^k$, $v(j, k) = w^j + w^k$, $v(i, j, k) = c$ for some

c satisfying $w^k < c < w^i + w^j + w^k$, and $v(S) = 0$ otherwise. Let (x^i, x^j, x^k) be its ϕ-solution. Consider the reduced games v_{-i}, v_{-j}, v_{-k}:

$$v_{-i}(j, k) = v(i, j, k) - \phi^i(\{i, j, k\}, v) = v(i, j, k) - x^i = x^j + x^k,$$

$$v_{-i}(j) = v(i, j) - \phi^i(\{i, j\}, v) = 0$$

(since $(\{i, j\}, v)$ is the null game), and

$$v_{-i}(k) = v(i, k) - \phi^i(\{i, k\}, v) = \phi^k(\{i, k\}, v)$$

$$= 0 + \frac{w^k}{w^k + w^i}[(w^i + w^k) - 0 - 0] = w^k$$

(recall (A.3) and Lemma A.6). Similarly:

$$v_{-j}(i, k) = x^i + x^k,$$

$$v_{-j}(i) = 0,$$

$$v_{-j}(k) = w^i.$$

Finally:

$$v_{-k}(i, j) = x^i + x^j,$$

$$v_{-k}(i) = v(i, k) - \phi^k(\{i, k\}, v) = w^i,$$

$$v_{-k}(j) = v(j, k) - \phi^k(\{j, k\}, v) = w^j.$$

Assume $x^k \leqslant w^k$. By (A.5) applied to v_{-i} we obtain $x^j \leqslant 0$; and by considering v_{-j}, we obtain $x^i < 0$. But this contradicts $x^i + x^j + x^k = c > w^k$. Therefore $x^k > w^k$, $x^j > 0$, $x^i > 0$, and moreover

$$\frac{x^j - 0}{x^k - w^k} = \frac{\alpha^j_{\{j,k\}}}{\alpha^k_{\{j,k\}}} = \frac{w^j}{w^k} \quad \text{(from } v_{-i}\text{)},$$

$$\frac{x^i - 0}{x^k - w^k} = \frac{w^i}{w^k} \quad \text{(from } v_{-j}\text{)}.$$

This implies

(A.11) $$\frac{x^i}{x^j} = \frac{w^i}{w^j}.$$

Next, consider v_k. Since $x^k > w^k$ and $x^i + x^j + x^k = c < w^i + w^j + w^k$, we have $x^i + x^j < w^i + w^j$, and

$$\frac{x^i - w^i}{x^j - w^j} = \frac{\beta^i_{\{i,j\}}}{\beta^j_{\{i,j\}}} = \frac{u^i}{u^j}.$$

But (A.11) implies $(x^i - w^i)/(x^j - w^j) = w^i/w^j$; hence $w^i/w^j = u^i/u^j$, or $w^i/u^i = w^j/u^j$. This holds for all $i \neq j$, proving our claim. *Q.E.D.*

The three lemmata and (A.3) thus imply (5.6): ϕ is the w-proportional solution for two-person games, completing the proof of the theorem. *Q.E.D.*

References

Aumann, R. J., and M. Maschler (1985): "Game Theoretic Analysis of a Bankruptcy Problem from the Talmud," *Journal of Economic Theory,* 36, 195–213. Balinsky, M. L., and H. P. Young (1982): *Fair Representation.* New Haven, CT: Yale University Press.

Aumann, R. J., and L. S. Shapley (1974): *Values of Non-Atomic Games.* Princeton, NJ: Princeton University Press.

Billera, L. J., and D. C. Heath (1982): "Allocation of Shared Cost: A Set of Axioms Yielding a Unique Procedure," *Mathematics of Operations Research,* 7, 32–39.

D'Aspremont, C., A. Jacquemin, and J.-F. Mertens (1984): "A Measure of Aggregate Power in Organizations," CORE DP-8416.

Davis, M., and M. Maschler (1965): "The Kernel of a Cooperative Game," *Naval Research Logistics Quarterly,* 12, 223–259.

Harsanyi, J. C. (1959): "A Bargaining Model for the Cooperative n-Person Game," in *Contributions to the Theory of Games IV* (*Annals of Mathematics Studies* 40), ed. by A. W. Tucker and R. D. Luce. Princeton: Princeton University Press, pp. 325–355.

Harsanyi, J. C. (1963): "A Simplified Bargaining Model for the n-Person Cooperative Game," *International Economic Review,* 4, 194–220.

Hart, S., and A. Mas-Colell (1988): "The Potential of the Shapley Value," in *The Shapley Value, Essays in Honor of Lloyd S. Shapley,* ed. by A. E. Roth. Cambridge: Cambridge University Press, pp. 127–137.

Kalai, E., and D. Samet (1985): "Monotonic Solutions to General Cooperative Games," *Econometrica,* 53, 307–327.

——— (1987): "On Weighted Shapley Values," *International Journal of Game Theory,* 16, 205–222.

Lensberg, T. (1982): "Stability and the Nash Solution," mimeo, Norwegian School of Economics and Business Administration, Bergen. *Journal of Economic Theory,* forthcoming.

Maschler, M., and G. Owen (1986): "The Consistent Shapley Value for Hyperplane Games," mimeo, RM-74, The Hebrew University, Jerusalem.

Mertens, J.-F. (1980): "Values and Derivatives," *Mathematics of Operations Research,* 4, 521–552.

Mirman, L. J., and Y. Tauman (1982): "Demand Compatible Equitable Cost-Sharing Prices," *Mathematics of Operations Research,* 7, 40–56.

Moulin, H. (1985): "The Separability Axiom and Equal-Sharing Methods," *Journal of Economic Theory,* 36, 120–148.

Myerson, R. B. (1980): "Conference Structures and Fair Allocation Rules," *International Journal of Game Theory,* 9, 169–182.

Ostroy, J. M. (1984): "A Reformulation of the Marginal Productivity Theory of Distribution," *Econometrica,* 52, 599–630.

Owen, G. (1968): "A Note on the Shapley Value," *Management Science,* 14, 731–732.

Peleg, B. (1985): "An Axiomatization of the Core of Cooperative Games Without Side Payments," *Journal of Mathematical Economics,* 14, 203–214.

——— (1986): "On the Reduced Game Property and its Converse," *International Journal of Game Theory,* 15, 187–200.

Shapley, L. S. (1953a): "Additive and Non-Additive Set Functions," Ph.D. Thesis, Princeton University.

——— (1953b): *"A Value for n-Person Games," in: Contributions to the Theory of Games II* (*Annals of Mathematics Studies* 28), ed. by H. W. Kuhn and A. W. Tucker. Princeton: Princeton University Press, 307–317.

——— (1981): "Comments on R. D. Banker's 'Equity Considerations in Traditional Full Cost Allocation Practices: An Axiomatic Perspective,'" in *Joint Cost Allocations,* ed. by S. Moriarity. Norman, OK: University of Oklahoma, pp. 131–136.

Sobolev, A. I. (1975): "The Characterization of Optimality Principles in Cooperative Games by Functional Equations" (in Russian), in *Mathematical Methods in Social Sciences,* ed. by N. N. Vorobev., Vilnius, 94–151.

Thomson, W. (1984): "Monotonicity, Stability and Egalitarianism," *Mathematical Social Sciences,* 8, 15–28.

— 8 —

An Equivalence Theorem for a Bargaining Set*

First, a modification of the Aumann-Maschler Bargaining Set is proposed. Then it is shown, under conditions of generality similar to the Core Equivalence Theorem, that the Bargaining Set and the set of Walrasian allocations coincide.

1. Introduction

A weakness of the Core as a solution concept in economics and game theory is that it depends on the notion that when a coalition objects to a proposed allocation, i.e., engages on an improving move, it neglects to take into account the repercussions triggered by the move. See Greenberg (1986) for a recent discussion of this point and its connection to the ideas underlying the von Neumann–Morgenstern stable-set solution.

Aumann and Maschler (1964) proposed a solution concept, the Bargaining Set, which addresses the above issue. Their idea is that for an objection based on a coalition to be effective it must be justified by the absence of a counterobjection. Objections which admit counterobjections

Journal of Mathematical Economics 18 (1989), 129–139.

* My first debt is to L. Shapley. It was because of his talk at the April 1986 Stony Brook Workshop on the Equivalence Principle (organized by A. Neyman) that I realized the Bargaining Set provided the proper interpretation for a result I presented at the same conference. Conversations and correspondence with J. Gabszewicz, B. Grodal, M. Maschler, B. Peleg, H. Scarf, A. Sengupta, K. Sengupta, B. Shitovitz, and R. Vohra have since been most helpful. The research was done in 1985–86 during my stay at MSRI, Berkeley, and the Department of Economics at the University of California, Berkeley. Thanks should go to both institutions. Financial support from the Guggenheim fellowship is gratefully acknowledged. Research supported in part by NSF Grant DMS-8120790.

are frivolous and, therefore, disregarded. Counterobjections are defined as proposals which are improving for some other coalition but which, however, guarantee that any common member of the two coalitions will be as well off as with the objection. This is, of course, an imprecise description of the idea and, in fact, there is not a unique definition of the Bargaining Set. From the original Aumann–Maschler concept several useful variants have evolved. See Maschler (1976), Owen (1982) and Shubik (1983).

The Bargaining Set is larger than the Core: blocking is harder. Nonetheless in the mid-seventies Shapley and Shubik [see Shubik (1984, ch. 12), and Shapley and Shubik (1984)] and then Geanakoplos (1978) showed that for transferable utility, differentiable economies the equivalence principle holds: if the economy is large the Bargaining Set shrinks to the set of Walrasian allocations. Shapley and Shubik considered sequences of increasingly large economies while Geanakoplos studies directly the limit situation relying on techniques from non-standard analysis.

In this paper we first propose a simplification of the Bargaining Set. We then establish the equivalence theorem under a generality analogous to Aumann (1964). One advantage of our definition is that it does not depend on distinguished individual players and it is thus well defined in the continuum case. Our proof is not in essence more complex than Aumann's although it is different and it includes an existence argument. The key idea is a characterization of justified objections as those objections that, in a precise sense, can be price supported. Our proof puts together three central arguments of equilibrium analysis, namely those that underline, respectively, the proofs of the two fundamental theorems and of the existence of equilibrium.

It is possible that the proposed redefinition of the Bargaining Set be of more general interest. To assess this, what is required is to see which forms the Bargaining Set takes in general games or in economic environments where the equivalence principle does not apply, e.g., situations where the notion of Walrasian equilibrium is not defined or where the Core is empty. An interesting fact [see Shitovitz (1988)] is that the equality of Bargaining Set and Walrasian allocations need not hold in cases where there are atoms and the Core equals the set of Walrasian allocations [as in Shitovitz (1973)]. Thus for these economic models the Bargaining Set discriminates better between inherently competitive and non-competitive environments.

See section VI for some additional comments on general games.

2. Definitions and a Theorem

There are l commodities. Our set of agents is $I = [0, 1]$. Lebesgue measure on I is denoted λ. With P the space of continuous and strictly monotone preferences on R^l_+ our economy is a (measurable) map $t \mapsto (\succsim_t, \omega(t))$ into $P \times R^l_{++}$ such that $\int \omega \ll \infty$. This economy remains fixed for the rest of the paper.

An *allocation* is a $x: I \to R^l_+$ such that $\int x \leqq \int \omega$. The allocation x is *Walrasian* if there is a $p \in R^l$ such that, for a.e. $t \in I$,: $p \cdot x(t) \leqq p \cdot \omega(t)$ and $p \cdot v > p \cdot \omega(t)$ whenever $v \succ_t x(t)$.

Definition 1. The pair (S, y) where $S \subset I$ and $y: S \to R^l_+$, is an objection to (or against) the allocation x if:

(a) $\int_S y \leqq \int_S \omega$,
(b) $y(t) \succsim_t x(t)$ for all a.e. $t \in S$ and $\lambda\{t \in S: y(t) \succ_t x(t)\} > 0$.

Definition 2. Let (S, y) be an objection to the allocation x. The pair (T, z), where $T \subset I$ and $z: T \to R^l_+$ is a counterobjection to (S, y) *if*:

(a) $\int_T z \leqq \int_T \omega$,
(b) $\lambda(T) > 0$, and
(c) (i) $z(t) \succ_t y(t)$ for a.e. $t \in T \cap S$,
(ii) $z(t) \succ_t x(t)$ for a.e. $t \in T \backslash S$.

Definition 3. An objection (S, y) is said to be justified if there is no counterobjection to it. The Bargaining Set is the set of allocations against which there is not justified objection.

A Walrasian allocation necessarily belongs to the Bargaining Set since there can be no objection, justified or not, against it (just apply the standard argument showing that Walrasian allocations belong to the Core). The aim of this paper is to show that the converse is also true

Theorem 1. If there is no justified objection against the allocation x then x is Walrasian. Hence, an allocation x belongs to the Bargaining Set if and only if it is Walrasian.

Observe that even the weaker result: 'if there is no justified objection against x then x is Pareto optimal' requires a non-trivial proof.

A topic for further research is the asymptotic version of the above theorem. Note that our Bargaining Set is well defined for finite economies and that it can well be larger than the Core (see example section 6).

Although the hypotheses of the Theorems are standard in equilibrium theory they are stronger than what is needed for the Core equivalence theorem in two respects. First, we require strict monotonicity of preferences and strictly positive endowments. This is because establishing existence of a type of equilibrium is part of our proof. Second, we postulate complete, transitive preferences. The extent to which Theorem 1 remains valid if preferences are unordered has been investigated by Grodal (1986). Yamazaki (1989) has studied the case where consumption sets are not required to be convex.

Related concepts of Bargaining Sets for economies have been discussed by Vind (1986).

Remark 1. Because of strict monotonicity of preferences the definition of counterobjection can be weakened to just requiring strict preference for a positive measure subset of the counterobjection coalition. For the theorem we cannot do with less than this. In particular strict preference cannot simply be replaced by weak preference (as it will be clear from careful consideration of the example in Remark 6).

Remark 2. If the presence of weak preference in the definition of objection is thought unattractive Theorem 1 will still obtain with strict preference if the Aumann–Maschler concept of the 'leader' of the objection is reintroduced à la manner of Geanakoplos (1978). Let $\delta > 0$ be an arbitrarily small number. Suppose now that given an allocation x we define a δ-objection to be a triple (K, S, y) such that $K \subset S$, $\lambda(K) \leqq \delta$, and (S, y) is an objection with $y(t) \succ_t x(t)$ for a.e. $t \in S$. A counterobjection to (K, S, y) is a counterobjection (T, z) to (S, y) such that $T \cap K = \phi$. A δ-objection is justified if there is no counterobjection to it.

Let there be a justified objection (S, y) against x. For any $\delta > 0$ we can select a group $K \subset \{t \in S: y(t) \succ_t x(t)\}, \lambda(K) < \delta$, and transfer some of their goods to the members of $S \backslash K$. The result is a δ-objection (K, s, y') which satisfies $y'(t) \succ_t y(t)$ for a.e. $t \in S \backslash K$. But this implies that any counterobjection to (K, S, y') is also necessarily a counterobjection to (S, y). Since the latter is justified it follows that (K, S, y') is also. Therefore Theorem 1 yields: 'given any $\delta > 0$, if there is no justified δ-objection against the allocation x then x is Walrasian'.

3. Walrasian Objections

The central idea for the proof of Theorem 1 is a consideration of a special class of objections generated by means of prices.

Definition 4. The objection (S, y) to the allocation x is Walrasian if there is a price system $p \neq 0$ such that, for a.e. t:

(i) $p \cdot v \geqq p \cdot \omega(t)$ for $v \succsim_t y(t)$, $t \in S$
(ii) $p \cdot v \geqq p \cdot \omega(t)$ for $v \succsim_t y(t)$, $t \in I \backslash S$.

Thus, roughly speaking, Walrasian objections have a self selection property. Given a p (and neglecting indifference) S is formed by precisely those agents who would rather trade at the price vector p than get the consumption bundle assigned to them by x.

The theorem is obtained by combining the following two Propositions:

Proposition 1. Any Walrasian objection (S, y) to an allocation x is justified.

Proposition 2. If x is not a Walrasian allocation then there is a Walrasian objection against it.

The next Proposition, the converse of Proposition 1, shows that the concept of Walrasian objection is more than a technical tool.

Proposition 3. If (S, y) is a justified objection to an allocation x then it is also a Walrasian objection.

Remark 3. By taking $\omega = x$ one gets, as a Corollary of Proposition 2, the existence of a Walrasian allocation for the entire economy.

Remark 4. Proposition 3 makes clear that typically there will be few justified objections against an objectionable allocation. Indeed justified objections have to be Walrasian and, while those exist (Proposition 2), they are obtained by solving a supply and demand problem that typically will have a finite number of solutions. Note the contrast with Core theory. It is well known that under the conditions of Proposition 2 if allocation is not Walrasian then it can be improved upon in a great variety of manners.

Remark 5. The fact that justified objections must be Walrasian also helps to understand why we cannot strengthen the concept of objection by requiring strict preference for a.e. $t \in S$. Suppose we are in a type economy and we consider an allocation x satisfying the equal treatment property.

Then the Walrasian objection (S, y), with objecting price vector p, will itself satisfy the equal treatment property in the following sense: if t, t' are of the same type and $t \in S$, $t' \notin S$ then $y(t) \sim_t x(t)$. This is easy to verify. Note first that $p \cdot y(t) = p \cdot \omega(t)$ because y must be a Walrasian allocation for S (which implies also $p \gg 0$). Suppose now that $y(t) \succ_t x(t)$. Then $y(t) \succ_{t'} x(t')$ and so $p \cdot y(t) > p \cdot \omega(t') = p \cdot \omega(t)$ which is impossible. Hence $y(t) \sim_t x(t)$. We must conclude therefore that if a coalition with a justified objection includes only part of some type of agents (and this may be unavoidable; see next two remarks) then it is not possible for these agents to strictly improve at the objection.

Remark 6. Given an allocation x and a price vector p any agent t having a Walrasian demand at p strictly preferred (resp. dispreferred) to $x(t)$ will be part (resp. remain outside) of any group attempting a Walrasian objection sustainable by p. The only freedom left corresponds to the marginal consumers i who are indifferent between getting $x(t)$ or their Walrasian demands at p.

Remark 7. It may be instructive to discuss an Edgeworth Box example. In fig. 1 there are two equal-mass types and we are considering the non-Walrasian, equal treatment allocation x. The curves BD and AC represent the relevant regions of the offer curves of the two types. Recalling the content of the previous remark it is easy to convince oneself that the price vectors p capable of sustaining a Walrasian objection can only have one of the following forms:

(i) p lies strictly between p_1 and p_2 and is an overall Walrasian equilibrium price (i.e., the curves BD and AC cross at the allocation corresponding to p). Then the objecting coalition is the coalition of the whole.

(ii) $p = p_1$ and the marginal type, type 1, is on the long side, i.e., B is to the left of A.* Then the objecting coalition includes all the consumers of type 2 and a fraction of those of type 1.

(iii) $p = p_2$ and the marginal type, in this case type 2, is on the long side, i.e., C is to the left of D. Then the objecting coalition includes all the consumers of type 1 and a fraction of those of type 2.

* Corrects the journal publication.

It is clear that at least one of the cases (i)–(iii) must occur. In the figure it is a case (ii).

Remark 8. A Walrasian objection constitutes, in particular, a Walrasian allocation for the objecting coalition. That any non-Walrasian allocation could be improved upon (even strictly) by a coalition using an allocation Walrasian for the coalition had been proved before [in Mas-Colell (1985) Proposition 7.3.2, as an extension of a result of Townsend (1983) for the replica case]. The existence of a Walrasian objection is, however, a much stronger property. For example, going back to fig 1 it should be clear that

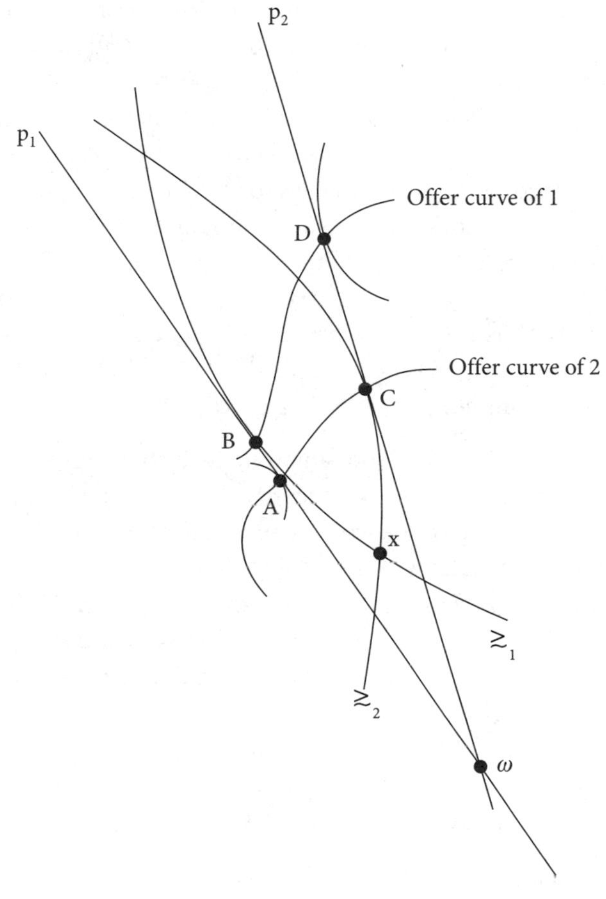

Figure 1

any price intermediate between p_1 and p_2 can be used to generate a coalition improving upon at one of its own Walrasian allocations. If, as a matter of definition, we restricted counterobjecting groups to be subsets of the corresponding objecting groups then Theorem 1 remains valid (objecting is now easier!) and justified objections would be characterized by being Walrasian allocations for the objecting coalition.

4. Proof of the Propositions

Proof of Proposition 1. The proof is just a repetition of the familiar argument establishing the optimality of equilibrium allocations (i.e., the *first fundamental theorem*).

Let p be the price vector associated with the Walrasian objection (S, y). Because $\int_S y \leqq \int_S \omega$ and $p \cdot y(t) \geqq p \cdot \omega(t)$ for a.e. $t \in S$ we have that y is a Walrasian allocation for S with price vector p. This also implies that $p \gg 0$ [remember that preferences are strictly monotone and $\omega(t) \gg 0$]. Suppose there is a counterobjection (T, z) to (S, y). Then for a.e. $t \in T \cap S$ we have $z(t) \succsim_t y(t)$ and therefore $p \cdot z(t) \geqq p \cdot y(t) \geqq p \cdot \omega(t)$, with strict inequality if $z(t) \succ_t y(t)$. For a.e. $t \in T \backslash S$ we have $z(t) \succsim_t x(t)$ and therefore $p \cdot z(t) \geqq p \cdot \omega(t)$, with strict inequality if $z(t) \succ_t x(t)$. We conclude that $\int_T p \cdot z(t) > \int_T p \cdot \omega(t)$ which contradicts $\int_T z \leqq \int_T \omega$. ■

Proof of Proposition 2. The proof amounts to an adaptation of the familiar arguments establishing the *existence of equilibrium* in economies with a continuum of traders and non-convex preferences. It is analogous to the proof of Proposition 7.3.2 in Mas-Colell (1985).

Let x be a allocation. For every $p \geqq 0$, $p \neq 0$, denote $C(p) = \{t$: there is v such that $v \succsim_t x(t)$ and $p \cdot v < p \cdot \omega(t)\}$. Observe that if $\lambda(C(p)) = 0$ for some p then x is Walrasian. Hence we assume from now on that $\lambda(C(p)) > 0$ for all p.

Denote $B = \{p \in R^l: ||p|| = 1, p \gg 0\}$ and let $f: B \times I \to R^l$ be the excess demand correspondence generated by our economy. We now define modified excess demand correspondence $f^*: B \times I \to R^l$ by

$$\begin{aligned} f^*(p,t) &= f(p,t) && \text{if } f(p,t) + \omega(t) \succ_t x(t), \\ &= f(p,t) \cup \{0\} && \text{if } f(p,t) + \omega(t) \sim_t x(t), \\ &= \{0\} && \text{if } f(p,t) + \omega(t) \prec_t x(t). \end{aligned}$$

Finally, put $F^*(p) = \int f^*(p, t)\, dt$. The correspondence $F^*: B \to R^l$ satisfies the following standard properties: (i) $p \cdot F^*(p) = 0$ for every p, (ii) F^* is upper hemicontinuous and bounded below, (iii) $F^*(p)$ is non-empty and convex for every p, and (iv) if $p_m \to p$, $p^j = 0$ for some j, and $v_m \in F^*(p_m)$ then $\|v_m\| \to \infty$. All this is easily verified; see Mas-Colell (1985, p. 270) for references to the technical facts. Note that the proof of (iii) requires a convexifying result (the Lyapunov–Richter Theorem) even if preferences are convex. Property (iv) follows from strict monotonicity and $\lambda(C(p)) > 0$.

The aggregate excess demand correspondence F^* satisfies, therefore, all the conditions for the existence of an equilibrium, i.e., there is $p \gg 0$ such that $F^*(p) = 0$ [see, e.g., Debreu (1970) for the argument]. Let $w: I \to R^l$ be such that $w(t) \in f^*(p, t)$ for all $t \in I$ and $\int w = 0$. Take $S = \{t: w(t) \in f(p, t)\}$ and define $y: S \to R^l_+$ by $y(t) = w(t) + \omega(t)$. We claim that (S, y) is a Walrasian objection to x with the price vector p. Indeed: (i) $\lambda(S) > 0$ because $C(p) \subset S$, (ii) $\int_S y \leqq \int_S \omega$ because $\int \omega = 0$ and, also, $w(t) = 0$ whenever $t \notin S$, (iii) if $t \in S$ and $v \succsim_t y(t)$ then $v \succsim_t f(p, t) + \omega(t)$ and so $p \cdot v \geqq p \cdot \omega(t)$, (iv) if $t \in I \backslash S$ then $x(t) \succsim_t f(p, t) + \omega(t)$ and so $v \succsim_t x(t)$ implies $p \cdot v \geqq p \cdot \omega(t)$. ■

Proof of Proposition 3. It amounts to a variation of the familiar Schmeidler's proof of the Aumann equivalence theorem [see, e.g., Hildenbrand (1974)] which can in turn be viewed as a sophisticated version of the *second fundamental theorem.*

Define $w: I \to R^l_+$ by $w(t) = y(t)$ if $t \in S$ and $w(t) = x(t)$ if $t \notin S$. Let then $V(t) = \{v - \omega(t): v \succeq_t w(t)\} \cup \{0\}$ and $V = \int V(t)\, dt$. By Lyapunov's theorem [e.g., L.1.3. in Mas-Colell (1985)] the set V is convex.

If $V \cap (-R^l_{++}) \neq \phi$ then there would be a counterobjection to (S, y). Therefore we can assume that $0 \notin \int S$. If we now let $p \neq 0$ support V at 0 we have that, for a.e. t, $p \cdot v \geqq p \cdot w(t)$ whenever $v \succsim_t w(t)$ which, of course, means that (S, y) is a Walrasian objection. ■

5. A Refinement

This section is in the spirit of an extended remark. The following definition iterates one more step the objection–counterobjection logic.

Definition 5. Let (T, z) be a counterobjection to the objection (S, y) to an allocation x. We say that (T, z) is justified if there is no (V, v) such that $\lambda(V) > 0$, $\int_V v \leqq \int_V \omega$ and, for $t \in V$,:

$$\begin{aligned} v(t) &\succ_t z(t) \quad \text{if} \quad t \in V \cap T, \\ v(t) &\succ_t y(t) \quad \text{if} \quad t \in V \cap S, \\ v(t) &\succ_t z(t) \quad \text{if} \quad t \in V \setminus (S \cup T).^{*} \end{aligned}$$

Suppose now that we modify our definition of counterobjection in section 2 to allow for weak preference (except for a positive measure subsets of T). Then we can strengthen our results by showing that any objection which admits a counterobjection (i.e., any non-Walrasian objection) admits in fact a justified counterobjection. The key is the following concept of Walrasian counterobjection:

Definition 6. The counterobjection (T, z) to the objection (S, y) to the allocation x is Walrasian if there is a price $p \neq 0$ such that, for a.e. t,:

(i) $p \cdot v \geqq p \cdot \omega(t)$ for $v \succsim_t z(t)$, $t \in T$,
(ii) $p \cdot v \geqq p \cdot \omega(t)$ for $v \succsim_t y(t)$, $t \in S$,
(iii) $p \cdot v \geqq p \cdot \omega(t)$ for $v \succsim_t x(t)$, $t \in I \setminus (S \cup T)$.

The proofs of Propositions 1 and 2 can then be easily adapted to show that, respectively: (i) *a Walrasian counterobjection is justified,* and (ii) *if an objection is not Walrasian* (hence, by Proposition 1, not justified) *then there is a Walrasian counterobjection against it.* In the proofs of the two Propositions we only have to replace x by a $x' : I \to R^l_+$ defined by $x'(t) = y(t)$ if $t \in S$ and $x'(t) = x(t)$ $t \notin S$. Substitute also the term objection by counterobjection and the symbols S, y, T, z by, respectively, T, z, V, v.

In a recent interesting contribution Dutta et al. (1987) have pushed a logic similar to this section to its limit and defined a Consistent Bargaining Set. There is little doubt that this is conceptually right and therefore we want to emphasize that none of these refinements and extensions detract from the strength of the Equivalence Theorem. On the contrary, by making counterobjecting easier we only make it more difficult for the equivalence to hold!

6. Comments on General Games

Our Bargaining Set concept is well defined for general games. Suppose that I is a (finite) set of players and $V: 2^I \to 2^{R^I}$ the characteristic form of a co-

* Corrects the journal publication.

operative game. Here $V(S) \subset R^I$ is the set of payoffs that coalition S can attain by itself.

Suppose that $x \in V(N)$. We say that (S, y) is an *objection* to x if $y \in V(S)$ and $y_t \geqq x_t$ for all $t \in S$, where at least one of the inequalities must be strict. A *counterobjection* to (S, y) is then a (T, z) such that $z \in V(T)$ and $z_t \geqq y_t$ (resp. $z_t \geqq x_t$) for $t \in T \cap S$ (resp. $t \in T \backslash S$). At least one of the inequalities must be strict. An objection is *justified* if it has no counterobjection. A $x \in V(I)$ is an *imputation* if it is weakly optimal (i.e., $x' \gg x$ implies $x' \notin V(I)$). The *Bargaining Set* if formed by the set of imputations against which there is no justified objection (important note: we do not require that imputations be individually rational).

Except for the addition of weak optimality (which disposes of some triviality) the above is the exact analog of the definition given for economies in section 2. We know very little about this Bargaining Set. For transferable utility games it is not difficult to verify that it contains the pre-Kernel of the game (the pre-Kernel has the same definition than the Kernel —see Shubik (1983)—except that individual rationality is not required. It contains the prenucleous and it is always non-empty) and it is thus non-empty. For non-superadditive general games the set may be empty [Peleg (1986)]. No counterexample to existence is yet available for the superadditive case. In fact it appears that in many examples the really serious problem is that the set can be quite large. Setting these issues seems particularly important. Progress has recently been made by Vohra (1989) and by Dutta et al. (1987) where, following the cue of Maschler et al. (1972), it is shown that for ordinally convex games the Bargaining Set is no larger than the Core.

We now give a transferable utility example showing that for totally balanced games (with at least four persons) our Bargaining Set may indeed be larger than the Core. This is of special interest to us because totally balanced games can be generated from exchange economies fitting the framework of section 2 [see Shapley and Shubik (1969)].

Example. $I = \{1, 2, 3, 4\}$ The game $v: 2^I \rightarrow R$ is normalized to $v(t) = 0$ for all $t \in I$ and is the minimal superadditive game compatible with the values $v(1234) = 4$, $v(123) = 3.1$ and $v(24) = v(34) = 2.06$. The game is totally balanced (a Core imputation is $(0, 1.9, 1.9, 0.2)$). The imputation $(1, 1, 1, 1)$ does not belong to the Core since it can be improved upon by several coalitions. It can also be seen that it does not belong to the Bargaining Set used by Shapley and Shubik [see Shubik (1984)]. Nonetheless, in our sense every objection against it admits a counterobjection and it thus belongs to our Bargaining Set.

What is the relation of our definition of the Bargaining Set to previous ones? Our Bargaining Set contains the one used by Shapley and Shubik in Shubik (1984). Strictly speaking the relation to the Aumann–Maschler Bargaining Set is one of non-comparability. This is because on the one hand we make counterobjecting easier (no member of the objection coalition is excluded a priori from belonging to a counterobjecting group) but on the other we make it a bit harder by requiring that at least one of the inequalities defining the counterobjection be strict.

References

Aumann, R., 1964, Markets with a continuum of traders, Econometrica 32, 39–50.

Aumann, R. and M. Maschler, 1964, The bargaining set for cooperative games, in: M. Dresher, L. S. Shapley and A. W. Tucker, eds., Advances in game theory (Princeton University Press, Princeton, NJ), 443–447.

Debreu, G., 1970, Economies with a finite set of equilibria, Econometrica 38, 387–392.

Dutta, B., D. Ray, K. Sengupta and R. Vohra, 1987, A consistent bargaining set, Journal of Economic Theory, forthcoming.

Geanakoplos, J., 1978, The bargaining set and nonstandard analysis, ch. 3 of Ph.D. dissertation (Harvard University, Cambridge, MA).

Greenberg, J., 1986, Stable standards of behavior: A unifying approach to solution concepts, Mimeo. (Stanford University, Stanford, CA).

Grodal, B., 1986, Bargaining sets and Walrasian allocations for atomless economies with incomplete preferences (MSRI, Berkeley, CA).

Hildenbrand, W., 1974, Core and equilibria of a large economy (Princeton University Press, Princeton, NJ).

Maschler, M., 1976, An advantage of the bargaining set over the core, Journal of Economic Theory 13, 184–194.

Maschler, M., B. Peleg and L. Shapley, 1972, The kernel and bargaining sets for convex games, International Journal of Game Theory 1, 73–93.

Mas-Colell, A., 1985, The theory of general economic equilibrium: A differentiable approach (Cambridge University Press, Cambridge).

Owen, G., 1982, Game theory (Academic Press, New York).

Peleg, B., 1986, Private communication.

Shapley, L. and M. Shubik, 1969, On market games, Journal of Economic Theory 1, 9–25.

Shapley, L. and M. Shubik, 1984, Convergence of the bargaining set for differentiable market games, Appendix B in Shubik, 1984, 683–692.

Shitovitz, B., 1973, Oligopoly in markets with a continuum of traders, Econometrica 41, 467–501.

Shitovitz, B., 1988, The bargaining set and the core in mixed markets with atoms and an atomless sector, Journal of Mathematical Economics, forthcoming.

Shubik, M., 1983, Game theory in the social sciences (MIT Press, Cambridge, MA).

Shubik, M., 1984, A game theoretic approach to political economy (MIT Press, Cambridge, MA).

Townsend, R., 1983, Theories of intermediated structures, in: K. Brunner and K. Meltzer, eds., Conference Series on Public Policy, Vol. 18, 221–272.

Vind, K., 1986, Two characterizations of bargaining sets, manuscript (MSRI, Berkeley, CA).

Vohra, R., 1989, An existence theorem for a bargaining set (Brown University, Providence, RI).

Yamazaki, A., 1989, Equilibria and bargaining sets without convexity assumptions, Discussion paper (Hitotsubashi University, Tokyo).

— 9 —

A Simple Adaptive Procedure Leading to Correlated Equilibrium[1]

SERGIU HART

ANDREU MAS-COLELL[2]

We propose a new and simple adaptive procedure for playing a game: "regret-matching." In this procedure, players may depart from their current play with probabilities that are proportional to measures of regret for not having used other strategies in the past. It is shown that our adaptive procedure guarantees that, with probability one, the empirical distributions of play converge to the set of correlated equilibria of the game.

Keywords: Adaptive procedure, correlated equilibrium, no regret, regret-matching, simple strategies.

1. Introduction

The leading noncooperative equilibrium notions for N-person games in strategic (normal) form are Nash equilibrium (and its refinements) and correlated equilibrium. In this paper we focus on the concept of correlated equilibrium.

A *correlated equilibrium*—a notion introduced by Aumann (1974)—can be described as follows: Assume that, before the game is played, each

Econometrica 68 (2000), 1127–1150.

1. October 1998 (minor corrections: June 1999). Previous versions: February 1998; November 1997; December 1996; March 1996 (handout). Research partially supported by grants of the U.S.-Israel Binational Science Foundation, the Israel Academy of Sciences and Humanities, the Spanish Ministry of Education, and the Generalitat de Catalunya.

2. We want to acknowledge the useful comments and suggestions of Robert Aumann, Antonio Cabrales, Dean Foster, David Levine, Alvin Roth, Reinhard Selten, Sylvain Sorin, an editor, the anonymous referees, and the participants at various seminars where this work was presented.

player receives a private signal (which does not affect the payoffs). The player may then choose his action in the game depending on this signal. A correlated equilibrium of the original game is just a Nash equilibrium of the game with the signals. Considering all possible signal structures generates all correlated equilibria. If the signals are (stochastically) independent across the players, it is a Nash equilibrium (in mixed or pure strategies) of the original game. But the signals could well be correlated, in which case new equilibria may obtain.

Equivalently, a correlated equilibrium is a probability distribution on N-tuples of actions, which can be interpreted as the distribution of play instructions given to the players by some "device" or "referee." Each player is given—privately—instructions for his own play only; the joint distribution is known to all of them. Also, for every possible instruction that a player receives, the player realizes that the instruction provides a best response to the random estimated play of the other players—assuming they all follow their instructions.

There is much to be said for correlated equilibrium. See Aumann (1974, 1987) for an analysis and foundational arguments in terms of rationality. Also, from a practical point of view, it could be argued that correlated equilibrium may be the most relevant noncooperative solution concept. Indeed, with the possible exception of well-controlled environments, it is hard to exclude a priori the possibility that correlating signals are amply available to the players, and thus find their way into the equilibrium.

This paper is concerned with dynamic considerations. We pose the following question: *Are there simple adaptive procedures always leading to correlated equilibrium?*

Foster and Vohra (1997) have obtained a procedure converging to the set of correlated equilibria. The work of Fudenberg and Levine (1999) led to a second one. We introduce here a procedure that we view as particularly simple and intuitive (see Section 4 for a comparative discussion of all these procedures). It does not entail any sophisticated updating, prediction, or fully rational behavior. Our procedure takes place in discrete time and it specifies that players adjust strategies probabilistically. This adjustment is guided by "regret measures" based on observation of past periods. Players know the past history of play of all players, as well as their own payoff matrix (but not necessarily the payoff matrices of the other players). Our Main Theorem is: The adaptive procedure generates trajectories of play that almost surely converge to the set of correlated equilibria.

The procedure is as follows: At each period, a player may either continue playing the same strategy as in the previous period, or switch to other strategies, with probabilities that are proportional to how much higher his accumulated payoff would have been had he always made that change in the past. More precisely, let U be his total payoff up to now; for each strategy k different from his last period strategy j, let $V(k)$ be the total payoff he would have received if he had played k every time in the past that he chose j (and everything else remained unchanged). Then only those strategies k with $V(k)$ larger than U may be switched to, with probabilities that are proportional to the differences $V(k) - U$, which we call the "regret" for having played j rather than k. These probabilities are normalized by a fixed factor, so that they add up to strictly less than 1; with the remaining probability, the same strategy j is chosen as in the last period.

It is worthwhile to point out three properties of our procedure. First, its simplicity; indeed, it is very easy to explain and to implement. It is not more involved than fictitious play (Brown (1951) and Robinson (1951); note that in the two-person zero-sum case, our procedure also yields the minimax value). Second, the procedure is *not* of the "best-reply" variety (such as fictitious play, smooth fictitious play (Fudenberg and Levine (1995, 1999)) or calibrated learning (Foster and Vohra (1997)); see Section 4 for further details). Players do not choose only their "best" actions, nor do they give probability close to 1 to these choices. Instead, all "better" actions may be chosen, with probabilities that are proportional to the apparent gains, as measured by the regrets; the procedure could thus be called *"regret-matching."* And third, there is "inertia." The strategy played in the last period matters: There is always a positive probability of continuing to play this strategy and, moreover, changes from it occur only if there is reason to do so.

At this point a question may arise: Can one actually guarantee that the smaller set of Nash equilibria is always reached? The answer is definitely "no." On the one hand, in our procedure, as in most others, there is a natural coordination device: the common history, observed by all players. It is thus reasonable to expect that, at the end, independence among the players will not obtain. On the other hand, the set of Nash equilibria is a mathematically complex set (a set of fixed-points; by comparison, the set of correlated equilibria is a convex polytope), and simple adaptive procedures cannot be expected to guarantee the global convergence to such a set.

After this introductory section, in Section 2 we present the model, describe the adaptive procedure, and state our result (the Main Theorem). Section 3 is devoted to a "stylized variation" of the procedure of Section 2. It is a variation that lends itself to a very direct proof, based on Blackwell's (1956a) Approachability Theorem. This is a new instrument in this field, which may well turn out to be widely applicable.

Section 4 contains a discussion of the literature, together with a number of relevant issues. The proof of the Main Theorem is relegated to the Appendix.

2. The Model and Main Result

Let $\Gamma = (N, (S^i)_{i \in N}, (u^i)_{i \in N})$ be a finite N-person game in strategic (normal) form: N is the set of players, S^i is the set of strategies of player i, and $u^i : \Pi_{i \in N} S^i \to \mathbb{R}$ is player i's payoff function. All sets N and S^i are assumed to be finite. Denote by $S := \Pi_{i \in N} S^i$ the set of N-tuples of strategies; the generic element of S is $s = (s^i)_{i \in N}$, and s^{-i} denotes the strategy combination of all players except i, i.e., $s^{-i} = (s^{i'})_{i' \neq i}$. We focus attention on the following solution concept:

Definition: A probability distribution ψ on S is a *correlated equilibrium* of Γ if, for every $i \in N$, every $j \in S^i$ and every $k \in S^i$ we have[3]

$$\sum_{s \in S: s^i = j} \psi(s)[u^i(k, s^{-i}) - u^i(s)] \leq 0.$$

If in the above inequality we replace the right-hand side by an $\varepsilon > 0$, then we obtain the concept of a *correlated ε-equilibrium.*

Note that every Nash equilibrium is a correlated equilibrium. Indeed, Nash equilibria correspond to the special case where ψ is a product measure, that is, the play of the different players is independent. Also, the set of correlated equilibria is nonempty, closed and convex, and even in simple games (e.g., "chicken") it may include distributions that are not in the convex hull of the Nash equilibrium distributions.

Suppose now that the game Γ is played repeatedly through time: $t = 1, 2, \ldots$. At time $t + 1$, given a history of play $h_t = (s_\tau)_{\tau=1}^t \in \Pi_{\tau=1}^t S$, we postulate that each player $i \in N$ chooses $s_{t+1}^i \in S^i$ according to a probability distribution[4] $p_{t+1}^i \in \Delta(S^i)$ which is defined in the following way:

3. We write $\Sigma_{s \in S: s^i = j}$ for the sum over all N-tuples s in S whose ith coordinate s^i equals j.
4. We write $\Delta(Q)$ for the set of probability distributions over a finite set Q.

For every two different strategies j, $k \in S^i$ of player i, suppose i were to replace strategy j, every time that it was played in the past, by strategy k; his payoff at time τ, for $\tau \leq t$, would become

$$W_\tau^i(j,k) := \begin{cases} u^i(k, s_\tau^{-i}), & \text{if } s_\tau^i = j, \\ u^i(s_\tau), & \text{otherwise.} \end{cases} \tag{2.1a}$$

The resulting difference in i's average payoff up to time t is then

$$\begin{aligned} D_t^i(j,k) &:= \frac{1}{t}\sum_{\tau=1}^{t} W_\tau^i(j,k) - \frac{1}{t}\sum_{\tau=1}^{t} u^i(s_\tau) \\ &= \frac{1}{t}\sum_{\tau\leq t:\, s_\tau^i=j} [u^i(k, s_\tau^{-i}) - u^i(s_\tau)]. \end{aligned} \tag{2.1b}$$

Finally, denote

$$R_t^i(j,k) := [D_t^i(j,k)]^+ = \max\{D_t^i(j,k), 0\}. \tag{2.1c}$$

The expression $R_t^i(j, k)$ has a clear interpretation as a measure of the (average) "regret" at period t for not having played, every time that j was placed in the past, the different strategy k.

Fix $\mu > 0$ to be a large enough number.[5] Let $j \in S^i$ be the strategy last chosen by player i, i.e., $j = s_t^i$. Then the probability distribution $p_{t+1}^i \in \Delta(S^i)$ used by i at time $t + 1$ is defined as

$$\begin{cases} p_{t+1}^i(k) := \dfrac{1}{\mu} R_t^i(j,k), & \text{for all } k \neq j, \\ p_{t+1}^i(k) := 1 - \displaystyle\sum_{k\in S^i:\, k\neq j} p_{t+1}^i(k). \end{cases} \tag{2.2}$$

Note that the choice of μ guarantees that $p_{t+1}^i(j) > 0$; that is, there is always a positive probability of playing the same strategy as in the previous period. The play $p_1^i \in \Delta(S^i)$ at the initial period is chosen arbitrarily.[6]

Informally, (2.2) may be described as follows. Player i starts from a "ref-

5. The parameter μ is fixed throughout the procedure (independent of time and history). It suffices to take μ so that $\mu > 2M^i(m^i - 1)$ for all $i \in N$, where M^i is an upper bound for $|u^i(\cdot)|$ and m^i is the number of strategies of player i. Even better, we could let μ satisfy $\mu > (m^i - 1)|\, u^i(k,s^{-i}) - u^i(j, s^{-i})|$ for all j, $k \in S^i$, all $s^{-i} \in S^{-i}$, and all $i \in N$ (and moreover we could use a different μ^i for each player i).

6. Actually, the procedure could start with any finite number of periods where the play is arbitrary.

erence point": his current actual play. His choice next period is governed by propensities to depart from it. It is natural therefore to postulate that, if a change occurs, it should be to actions that are perceived as being better, relative to the current choice. In addition, and in the spirit of adaptive behavior, we assume that all such better choices get positive probabilities; also, the better an alternative action seems, the higher the probability of choosing it next time. Further, there is also inertia: the probability of staying put (and playing the same action as in the last period) is always positive.

More precisely, the probabilities of switching to different strategies are proportional to their regrets relative to the current strategy. The factor of proportionality is constant. In particular, if the regrets are small, then the probability of switching from current play is also small.

For every t, let $z_t \in \Delta(S)$ be the empirical distribution of the N-tuples of strategies played up to time t. That is, for every[7] $s \in S$,

$$z_t(s) := \frac{1}{t}|\{\tau \leq t : s_\tau = s\}| \tag{2.3}$$

is the relative frequency that the N-tuple s has been played in the first t periods. We can now state our main result.

Main Theorem: If every player plays according to the adaptive procedure (2.2), *then the empirical distributions of play z_t converge almost surely as $t \to \infty$ to the set of correlated equilibrium distributions of the game Γ.*

Note that convergence to the *set* of correlated equilibria does not imply that the sequence z_t converges to *a point*. The Main Theorem asserts that the following statement holds with probability one: For any $\varepsilon > 0$ there is $T_0 = T_0(\varepsilon)$ such that for all $t > T_0$ we can find a correlated equilibrium distribution ψ_t at a distance less than ε from z_t. (Note that this T_0 depends on the history; it is an "a.s. finite stopping time.") That is, the Main Theorem says that, with probability one, for any $\varepsilon > 0$, the (random) trajectory $(z_1, z_2, \ldots, z_t, \ldots)$ enters and then stays forever in the ε-neighborhood in $\Delta(S)$ of the set of correlated equilibria. Put differently: Given any $\varepsilon > 0$, there exists a constant (i.e., independent of history) $t_0 = t_0(\varepsilon)$ such that, with probability at least $1 - \varepsilon$, the empirical distributions z_t for *all* $t > t_0$ are in the ε-neighborhood of the set of correlated equilibria. Finally, let us note that because the set of correlated equilibria is nonempty and com

7. We write $|Q|$ for the number of elements of a finite set Q.

pact, the statement "the trajectory (z_t) converges to the set of correlated equilibria" is equivalent to the statement "the trajectory (z_t) is such that for any $\varepsilon > 0$ there is $T_1 = T_1(\varepsilon)$ with the property that z_t is a correlated ε-equilibrium for all $t > T_1$."

We conclude this section with a few comments (see also Section 4):

(1) Our adaptive procedure (2.2) requires player i to know his own payoff matrix (but not those of the other players) and, at time $t + 1$, the history h_t; actually, the empirical distribution z_t of $(s_1, s_2, \ldots, s_t)$ suffices. In terms of computation, player i needs to keep record of the time t together with the $m^i(m^i - 1)$ numbers $D_t^i(j, k)$ for all $j \neq k$ in S^i (and update these numbers every period).

(2) At every period the adaptive procedure that we propose randomizes only over the strategies that exhibit positive regret relative to the most recently played strategy. Some strategies may, therefore, receive zero probability. Suppose that we were to allow for trembles. Specifically, suppose that at every period we put a $\delta > 0$ probability on the uniform tremble (each strategy thus being played with probability at least δ/m^i). It can be shown that in this case the empirical distributions z_t converge to the set of correlated ε-equilibria (of course, ε depends on δ, and it goes to zero as δ goes to zero). In conclusion, unlike most adaptive procedures, ours does not rely on trembles (which are usually needed, technically, to get the "ergodicity" properties); moreover, our result is robust with respect to trembles.

(3) Our adaptive procedure depends only on one parameter,[8] μ. This may be viewed as an "inertia" parameter (see Subsections 4(g) and 4(h)): A higher μ yields lower probabilities of switching. The convergence to the set of correlated equilibria is always guaranteed (for any large enough μ; see footnote 5), but the speed of convergence changes with μ.

(4) We know little about additional convergence properties for z_t. It is easy to see that the empirical distributions z_t either converge to a Nash equilibrium in pure strategies, or must be infinitely often outside the set of correlated equilibria (because, if z_t is a correlated equilibrium from some time on, then[9] all regrets are 0, and the play does not change). This implies, in particular, that interior (relative to $\Delta(S)$) points of the set of correlated equilibria that are not pure Nash equilibria are unreachable as the limit of some z_t (but it is possible that they are reachable as limits of a *subsequence* of z_t).

8. Using a parameter μ (rather than a fixed normalization of the payoffs) was suggested to us by Reinhard Selten.

9. See the Proposition in Section 3.

(5) There are other procedures enjoying convergence properties similar to ours: the procedures of Foster and Vohra (1997), of Fudenberg and Levine (1999), and of Theorem A in Section 3 below; see the discussion in Section 4. The delimitation of general classes of procedures converging to correlated equilibria seems, therefore, an interesting research problem.[10]

3. No Regret and Blackwell Approachability

In this section (which can be viewed as a motivational preliminary) we shall replace the adaptive procedure of Section 2 by another procedure that, while related to it, is more stylized. Then we shall analyze it by means of Blackwell's (1956a) Approachability Theorem, and prove that it yields convergence to the set of correlated equilibria. In fact, the Main Theorem stated in Section 2, and its proof in Appendix 1, were inspired by consideration and careful study of the result of this section. Furthermore, the procedure here is interesting in its own right (see, for instance, the Remark following the statement of Theorem A, and (d) in Section 4).

Fix a player i and recall the procedure of Section 2: At time $t+1$ the transition probabilities, from the strategy played by player i in period t to the strategies to be played at $t+1$, are determined by the stochastic matrix defined by the system (2.2). Consider now an invariant probability vector $q_t^i = (q_t^i(j))_{j \in S^i} \in \Delta(S^i)$ for this matrix (such a vector always exists). That is, q_t^i satisfies

$$q_t^i(j) = \sum_{k \neq j} q_t^i(k) \frac{1}{\mu} R_t^i(k, j) + q_t^i(j) \left[1 - \sum_{k \neq j} \frac{1}{\mu} R_t^i(j, k) \right],$$

for every $j \in S^i$. By collecting terms, multiplying by μ, and formally letting $R_t^i(j, j) := 0$, the above expression can be rewritten as

$$\sum_{k \in S^i} q_t^i(k) R_t^i(k, j) = q_t^i(j) \sum_{k \in S^i} R_t^i(j, k), \tag{3.1}$$

for every $j \in S^i$

In this section we shall assume that play at time $t+1$ by player i is determined by a solution q_t^i to the system of equations (3.1); i.e., $p_{t+1}^i(j) := q_t^i(j)$. In a sense, we assume that player i at time $t+1$ goes instantly to the invariant distribution of the stochastic transition matrix determined by (2.2). We now state the key result.

10. See Hart and Mas-Colell (1999) and Cahn (2000) for such results.

Theorem A: *Suppose that at every period* $t+1$ *player* i *chooses strategies according to a probability vector* q_t^i *that satisfies* (3.1). *Then player i's regrets* $R_t^i(j,k)$ *converge to zero almost surely for every* j, k *in* S^i *with* $j \neq k$.

Remark: Note that—in contrast to the Main Theorem, where every player uses (2.2)—no assumption is made in Theorem A on how players different from i choose their strategies (except for the fact that for every t, given the history up to t, play is independent among players). In the terminology of Fudenberg and Levine (1999, 1998), the adaptive procedure of this section is "(universally) calibrated." For an extended discussion of this issue, see Subsection 4(d).

What is the connection between regrets and correlated equilibria? It turns out that a necessary and sufficient condition for the empirical distributions to converge to the set of correlated equilibria is precisely that all regrets converge to zero. More generally, we have the following proposition.

Proposition: *Let* $(s_t)_{t=1,2,\ldots}$ *be a sequence of plays (i.e.,* $s_t \in S$ *for all* t*) and let*[11] $\varepsilon \geq 0$. *Then:* $\limsup_{t\to\infty} R_t^i(j,k) \leq \varepsilon$ *for every* $i \in N$ *and every* $j, k \in S^i$ *with* $j \neq k$, *if and only if the sequence of empirical distributions* z_t *(defined by* (2.3)*) converges to the set of correlated* ε*-equilibria.*

Proof: For each player i and every $j \neq k$ in S^i we have

$$D_t^i(j,k) = \frac{1}{t} \sum_{\tau \leq t : s_\tau^i = j} [u^i(k, s_\tau^{-i}) - u^i(j, s_\tau^{-i})]$$
$$= \sum_{s \in S : s^i = j} z_t(s)[u^i(k, s^{-i}) - u^i(j, s^{-i})].$$

On any subsequence where z_t converges, say $z_{t'} \to \psi \in \Delta(S)$, we get

$$D_{t'}^i(j,k) \to \sum_{s \in S : s^i = j} \psi(s)[u^i(k, s^{-i}) - u^i(j, s^{-i})].$$

The result is immediate from the definition of a correlated ε-equilibrium and (2.1c). *Q.E.D.*

Theorem A and the Proposition immediately imply the following corollary.

Corollary: *Suppose that at each period* $t+1$ *every player* i *chooses strategies according to a probability vector* q_t^i *that satisfies* (3.1). *Then the empiri-*

11. Note that both $\varepsilon > 0$ and $\varepsilon = 0$ are included.

cal distributions of play z_t converge almost surely as $t \to \infty$ to the set of correlated equilibria of the game Γ.

Before addressing the formal proof of Theorem A, we shall present and discuss Blackwell's Approachability Theorem.

The basic setup contemplates a decision-maker i with a (finite) action set S^i. For a finite indexing set L, the decision-maker receives an $|L|$-dimensional vector payoff $v(s^i, s^{-i}) \in \mathbb{R}^L$ that depends on his action $s^i \in S^i$ and on some external action s^{-i} belonging to a (finite) set S^{-i} (we will refer to $-i$ as the "opponent"). The decision problem is repeated through time. Let $s_t = (s_t^i, s_t^{-i}) \in S^i \times S^{-i}$ denote the choices at time t; of course, both i and $-i$ may use randomizations. The question is whether the decision-maker i can guarantee that the time average of the vector payoffs, $D_t := (1/t)\Sigma_{\tau \le t} v(s_T) \equiv (1/t)\Sigma_{\tau \le t} v\left(s_\tau^i, s_\tau^{-i}\right)$, approaches a predetermined set (in $\mathbb{R}^L$).

Let $\mathscr{C}$ be a convex and closed subset of $\mathbb{R}^L$. The set $\mathscr{C}$ is *approachable* by the decision-maker i if there is a procedure[12] for i that guarantees that the average vector payoff D_t approaches the set $\mathscr{C}$ (i.e.,[13] $\text{dist}(D_t, \mathscr{C}) \to 0$ almost surely as $t \to \infty$), regardless of the choices of the opponent $-i$. To state Blackwell's result, let $w_{\mathscr{C}}$ denote the *support function* of the convex set $\mathscr{C}$, i.e., $w_{\mathscr{C}}(\lambda) := \sup\{\lambda \cdot c : c \in \mathscr{C}\}$ for all λ in $\mathbb{R}^L$. Given a point $x \in \mathbb{R}^L$ which is not in $\mathscr{C}$, let $F(x)$ be the (unique) point in $\mathscr{C}$ that is closest to x in the Euclidean distance, and put $\lambda(x) := x - F(x)$; note that $\lambda(x)$ is an outward normal to the set $\mathscr{C}$ at the point $F(x)$.

Blackwell's Approachability Theorem: Let $\mathscr{C} \subset \mathbb{R}^L$ be a convex and closed set, with support function $w_{\mathscr{C}}$. Then $\mathscr{C}$ is approachable by i if and only if for every $\lambda \in \mathbb{R}^L$ there exists a mixed strategy $q_\lambda \in \Delta(S^i)$ such that[14]

$$\lambda \cdot v(q_\lambda, s^{-i}) \le w_{\mathscr{C}}(\lambda), \qquad \textit{for all } s^{-i} \in S^{-i}. \tag{3.2}$$

Moreover, the following procedure of i guarantees that $\text{dist}(D_t, \mathscr{C})$ *converges*

12. In the repeated setup, we refer to a (behavior) strategy as a "procedure."

13. $\text{dist}(x, A) := \min\{\|x - a\| : a \in A\}$, where $\|\cdot\|$ is the Euclidean norm. Strictly speaking, Blackwell's definition of approachability requires also that the convergence of the distance to 0 be uniform over the procedures of the opponent; i.e., there is a procedure of i such that for every $\varepsilon > 0$ there is $t_0 \equiv t_0(\varepsilon)$ such that for any procedure of $-i$ we have $P[\text{dist}(D_t, \mathscr{C}) < \varepsilon \text{ for all } t > t_0] > 1 - \varepsilon$. The Blackwell procedure (defined in the next Theorem) guarantees this as well.

14. $v(q, s^{-i})$ denotes the expected payoff, i.e., $\sum_{s^i \in S^i} q(s^i) v(s^i, s^{-i})$. Of course, only $\lambda \ne 0$ with $w_{\mathscr{C}}(\lambda) < \infty$ need to be considered in (3.2).

almost surely to 0 as $t \to \infty$: At time $t+1$, play $q_{\lambda(D_t)}$ if $D_t \notin \mathscr{C}$, and play arbitrarily if $D_t \in \mathscr{C}$.

We will refer to the condition for approachability given in the Theorem as the *Blackwell condition,* and to the procedure there as the *Blackwell procedure.* To get some intuition for the result, assume that D_t is not in $\mathscr{C}$, and let $\mathscr{H}(D_t)$ be the half-space of $\mathbb{R}^L$ that contains $\mathscr{C}$ (and not D_t) and is bounded by the supporting hyperplane to $\mathscr{C}$ at $F(D_t)$ with normal $\lambda(D_t)$; see Figure 1. When i uses the Blackwell procedure, it guarantees that $v(q_{\lambda(D_t)}, s^{-i})$ lies in $\mathscr{H}(D_t)$ for all s^{-i} in S^{-i} (by (3.2)). Therefore, given D_t, the expectation of the next period payoff $E[v(s_{t+1})|D_t]$ will lie in the half-space $\mathscr{H}(D_t)$ for any pure choice s_{t+1}^{-i} of $-i$ at time $t+1$, and thus also for any randomized choice of $-i$. The expected average vector payoff at period $t+1$ (conditional on D_t) is

$$E[D_{t+1}\,|\,D_t] = \frac{t}{t+1}D_t + \frac{1}{t+1}E[v(s_{t+1})\,|\,D_t].$$

When t is large, $E[D_{t+1}|D_t]$ will thus be *inside* the circle of center $F(D_t)$ and radius $||\lambda(D_t)||$. Hence

$$\mathrm{dist}(E[D_{t+1}\,|D_t], \mathscr{C}) \le ||E[D_{t+1}|D_t] - F(D_t)\,|| < ||\,\lambda(D_t)\,|| = \mathrm{dist}\,(D_t, \mathscr{C})$$

(the first inequality follows from the fact that $F(D_t)$ is in $\mathscr{C}$). A precise computation shows that the distance not only decreases, but actually goes to zero.[15] For proofs of Blackwell's Approachability Theory, see[16] Blackwell (1956a), or Mertens, Sorin, and Zamir (1995, Theorem 4.3).

We now prove Theorem A.

Proof of Theorem A: As mentioned, the proof of this Theorem consists of an application of Blackwell's Approachability Theorem. Let

$$L := \{(j, k) \in S^i \times S^i : j \neq k\},$$

15. Note that one looks here at *expected* average payoffs; the Strong Law of Large Numbers for Dependent Random Variables—see the Proof of Step M10 in the Appendix—implies that the *actual* average payoffs also converge to the set $\mathscr{C}$.

16. The Blackwell condition is usually stated as follows: For every $x \notin \mathscr{C}$ there exists $q(x) \in \Delta(S^i)$ such that $[x - F(x)]\cdot[v(q(x), s^{-i}) - F(x)] \le 0$, for all $s^{-i} \in S^{-i}$. It is easy to verify that this is equivalent to our formulation. We further note a simple way of stating the Blackwell result: A convex set $\mathscr{C}$ is approachable if and only if any half-space containing $\mathscr{C}$ is approachable.

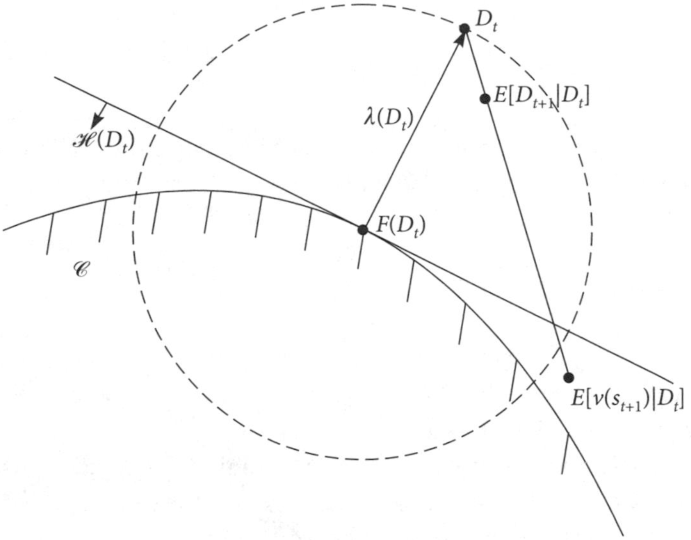

Figure 1. Approaching the set $\mathscr{C}$.

and define the vector payoff $v(s^i, s^{-i}) \in \mathbb{R}^L$ by letting its $(j, k) \in L$ coordinate be

$$[v(s^i, s^{-i})](j,k) := \begin{cases} u^i(k, s^{-i}) - u(j, s^{-i}), & \text{if } s^i = j, \\ 0, & \text{otherwise.} \end{cases}$$

Let $\mathscr{C}$ be the nonpositive orthant $\mathbb{R}^L_- := \{x \in \mathbb{R}^L : x \leq 0\}$. We claim that $\mathscr{C}$ is approachable by i. Indeed, the support function of $\mathscr{C}$ is given by $w_{\mathscr{C}}(\lambda) = 0$ for all $\lambda \in \mathbb{R}^L_+$ and $w_{\mathscr{C}}(\lambda) = \infty$ otherwise; so only $\lambda \in \mathbb{R}^L_+$ need to be considered. Condition (3.2) is

$$\sum_{(j,k) \in L} \lambda(j,k) \sum_{s^i \in S^i} q_\lambda(s^i)[v(s^i, s^{-i})](j,k) \leq 0,$$

or

(3.3) $$\sum_{(j,k) \in L} \lambda(j,k) q_\lambda(j)[u^i(k, s^{-i}) - u^i(j, s^{-i})] \leq 0$$

for all $s^{-i} \in S^{-i}$. After collecting terms, the left-hand side of (3.3) can be written as

(3.4a) $$\sum_{j \in S^i} \alpha(j) u^i(j, s^{-i}),$$

where

$$\alpha(j) := \sum_{k \in S^i} q_\lambda(k)\, \lambda(k, j) - q_\lambda(j) \sum_{k \in S^i} \lambda(j, k). \tag{3.4b}$$

Let $q_\lambda \in \Delta(S^i)$ be an invariant vector for the nonnegative $S^i \times S^i$ matrix with entries $\lambda(j, k)$ for $j \neq k$ and 0 for $j = k$ (such a q_λ always exists). That is, q_λ satisfies

$$\sum_{k \in S^i} q_\lambda(k)\, \lambda(k, j) = q_\lambda(j) \sum_{k \in S^i} \lambda(j, k), \tag{3.5}$$

for every $j \in S^i$. Therefore $\alpha(j) = 0$ for all $j \in S^i$, and so inequality (3.3) holds true (as an equality[17]) for all $s^{-i} \in S^{-i}$. The Blackwell condition is thus satisfied by the set $\mathscr{C} = \mathbb{R}^L_-$.

Consider D_t, the average payoff vector at time t. Its (j, k)-coordinate is $(1/t)\Sigma_{\tau \leq t}[v(s_\tau)](j, k) = D^i_t(j, k)$. If $D_t \notin \mathbb{R}^L_-$, then the closest point to D_t in $\mathbb{R}^L_-$ is $F(D_t) = [D_t]^-$ (see Figure 2), hence $\lambda(D_t) = D_t - [D_t]^- = [D_t]^+ = (R^i_t(j, k))_{(j,k) \in L}$, which is the vector of regrets at time t. Now the given strategy of i at time $t + 1$ satisfies (3.1), which is exactly condition (3.5) for $\lambda = \lambda(D_t)$. Hence player i uses the Blackwell procedure for $\mathbb{R}^L_-$, which guarantees that the average vector payoff D_t approaches $\mathbb{R}^L_-$, or $R^i_t(j, k) \to 0$ a.s. for every $j \neq k$. *Q.E.D.*

Remark: The proof of Blackwell's Approachability Theorem also provides bounds on the speed of convergence. In our case, one gets the following: The expectation $E[R^i_t(j, k)]$ of the regrets is of the order of $1/\sqrt{t}$, and the probability that z_t is a correlated ε-equilibrium for all $t > T$ is at least $1 - ce^{-cT}$ (for an appropriate constant $c > 0$ depending on ε; see Foster and Vohra (1999, Section 4.1)). Clearly, a better speed of convergence[18] for the expected regrets cannot be guaranteed, since, for instance, if the other players play stationary mixed strategies, then the errors are of the order $1/\sqrt{t}$ by the Central Limit Theorem.

4. Discussion

This section discusses a number of important issues, including links and comparisons to the relevant literature.

17. Note that this is precisely Formula (2) in the Proof of Theorem 1 in Hart and Schmeidler (1989); see Subsection 4(i).

18. Up to a constant factor.

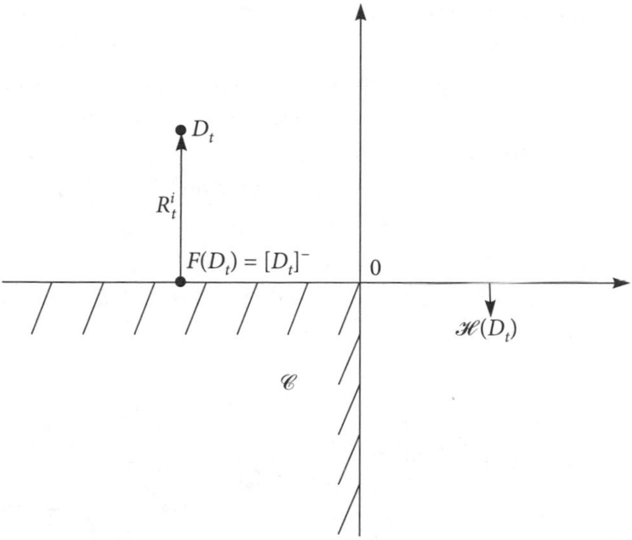

Figure 2. Approaching $\mathscr{C} = \mathbb{R}_-^L$.

(a) *Foster and Vohra.* The seminal paper in this field of research is Foster and Vohra (1997). They consider, first, "forecasting rules"—on the play of others—that enjoy good properties, namely, "calibration." Second, they assume that each player best-replies to such calibrated forecasts. The resulting procedure leads to correlated equilibria. The motivation and the formulation are quite different from ours; nonetheless, their results are close to our results (specifically, to our Theorem A), since their calibrated forecasts are also based on regret measures.[19]

(b) *Fudenberg and Levine.* The next important paper is Fudenberg and Levine (1999) (see also their book (1998)). In that paper they offer a class of adaptive procedures, called "calibrated smooth fictitious play," with the property that for every $\varepsilon > 0$ there are procedures in the class that guarantee almost sure convergence to the set of correlated ε-equilibria (but the conclusion does not hold for $\varepsilon = 0$). The formal structure of these procedures is also similar to that of our Theorem A, in the sense that the mixed choice of a given player at time t is determined as an invariant probability

19 These regrets are defined on an ε-grid on $\Delta(S^{-i})$, with ε going to zero as t goes to infinity. Therefore, at each step in their procedure one needs to compute the invariant vector for a matrix of an increasingly large size; by comparison, in our Theorem A the size of the matrix is fixed, $m^i \times m^i$.

vector of a transition matrix. However, the transition matrix (and therefore the stochastic dynamics) is different from the regret-based transition matrix of our Theorem A. To understand further the similarities and differences between the Fudenberg and Levine procedures and our own, the next two Subsections, (c) and (d), contain a detour on the concepts of "universal consistency" and "universal calibration."

(c) *Universal Consistency.* The term "universal consistency" is due to Fudenberg and Levine (1995). The concept goes back to Hannan (1957), who proved the following result: There is a procedure (in the setup of Section 2) for player i that guarantees, no matter what the other players do, that

$$\limsup_{t\to\infty}\left[\max_{k\in S^i}\frac{1}{t}\sum_{\tau=1}^{t}u^i(k,s_\tau^{-i})-\frac{1}{t}\sum_{\tau=1}^{t}u^i(s_\tau)\right]\le 0 \qquad \text{a.s.} \tag{4.1}$$

In other words, i's average payoff is, in the limit, no worse than if he were to play any *constant* strategy $k \in S^i$ for all $\tau \le t$. This property of the Hannan procedure for player i is called *universal consistency* by Fudenberg and Levine (1995) (it is "universal" since it holds no matter how the other players play). Another universally consistent procedure was shown by Blackwell (1956b) to result from his Approachability Theorem (see also Luce and Raiffa (1957, pp. 482–483)).

The adaptive procedure of our Theorem A is also universally consistent. Indeed, for each j in S^i, (4.1) is guaranteed even when restricted to those periods when player i chose that particular j; this being true for all j in S^i, the result follows. However, the application of Blackwell's Approachability Theorem in Section 3 suggests the following particularly simple procedure.

At time t, for each strategy k in S^i, let

$$D_t^i(k) := \frac{1}{t}\sum_{\tau=1}^{t}[u^i(k,s_\tau^{-i})-u^i(s_\tau)], \tag{4.2a}$$

$$p_{t+1}^i(k) := \frac{[D_t^i(k)]^+}{\sum_{k'\in S^i}[D_t^i(k')]^+}, \tag{4.2b}$$

if the denominator is positive, and let $p_{t+1}^i \in \Delta(S^i)$ be arbitrary otherwise. The strategy of player i is then, at time $t + 1$, to choose k in S^i with probability $p_{t+1}^i(k)$. These probabilities are thus proportional to the "unconditional regrets" $[D_t^i(k)]^+$ (by comparison to the "conditional on j" regrets of Section 2). We then have the following theorem.

Theorem B: The adaptive procedure (4.2) *is universally consistent for player i.*

The proof of Theorem B is similar to the proof of Theorem A in Section 3 and is omitted.

Fudenberg and Levine (1995) propose a class of procedures that turn out to be universally ε-consistent:[20] "smooth fictitious play." Player i follows a smooth fictitious play behavior rule if at time t he plays a mixed strategy $\sigma^i \in \Delta(S^i)$ that maximizes the sum of his expected payoff (with the actions of the remaining players distributed as in the empirical distribution up to t) and $\lambda v^i(\sigma^i)$, where $\lambda > 0$ and v^i is a strictly concave smooth function defined on i's strategy simplex, $\Delta(S^i)$, with infinite length gradient at the boundary of $\Delta(S^i)$. The result of Fudenberg and Levine is then that, given any $\varepsilon > 0$, there is a sufficiently small λ such that universal ε-consistency obtains for player i. Observe that, for small λ, smooth fictitious play is very close to fictitious play (it amounts to playing the best response with high probability and the remaining strategies with low but positive probability). The procedure is, therefore, clearly distinct from (4.2): In (4.2) all the better, even if not best, replies are played with significant probability; also, in (4.2) the inferior replies get zero probability. Finally, it is worth emphasizing that the tremble from best response is required for the Fudenberg and Levine result, since fictitious play is not guaranteed to be consistent. In contrast, the procedure of (4.2) has no trembles.

The reader is referred to Hart and Mas-Colell (1999), where a wide class of universally consistent procedures is exhibited and characterized (including as special cases (4.2) as well as smooth fictitious play).

(d) *Universal Calibration.* The idea of "universal calibration," also introduced[21] by Fudenberg and Levine (1998, 1999), is that, again, regret measures go to zero irrespective of the other players' play. The difference is that, now, the set of regret measures is richer: It consists of regrets that are conditional on the strategy currently played by i himself. Recall the Proposition of Section 3: If such universally calibrated strategies are played by all

20. That is, the right-hand side of (4.1) is $\varepsilon > 0$ instead of 0.

21. They actually call it "calibration"; we prefer the term "universal calibration," since it refers to *any* behavior of the opponents (as in their "[conditional] universal consistency").

players, then all regrets become nonpositive in the limit, and thus the convergence to the correlated equilibrium set is guaranteed.

The procedure of Theorem A is universally calibrated; so (up to ε) is the "calibrated smooth fictitious play" of Fudenberg and Levine (1999). The two procedures stand to each other as, in the unconditional version, Theorem B stands to "smooth fictitious play."

The procedure (2.2) of our Main Theorem is not universally calibrated. If only player i follows the procedure, we cannot conclude that all his regrets go to zero; adversaries who know the procedure used by player i could keep his regrets positive.[22] Such sophisticated strategies of the other players, however, are outside the framework of our study—which deals with simple adaptive behavior. In fact, it turns out that the procedure of our Main Theorem is guaranteed to be calibrated not just against opponents using the same procedure, but also against a wide class of behaviors.[23]

We regard the simplicity of (2.2) as a salient point. Of course, if one needs to guarantee calibration even against sophisticated adversaries, one may have to give up on simplicity and resort to the procedure of Theorem A instead.

(e) *Better-reply vs. Best-reply.* Note that all the procedures in the literature reviewed above are best-reply-based: A player uses (almost) exclusively actions that are (almost) best-replies to a certain belief about his opponents. In contrast, our procedure gives significant probabilities to any actions that are just better (rather than best). This has the additional effect of making the behavior continuous, without need for approximations.

(f) *Eigenvector Procedures.* The procedure of our Main Theorem differs from all the other procedures leading to correlated equilibria (including that of our Theorem A) in an important aspect: It does not require the player to compute, at every step, an invariant (eigen-) vector for an appropriate positive matrix. Again, the simplicity[24] of (2.2) is an essential prop-

22. At each time $t+1$, let them play an $(N-1)$-tuple of strategies that minimizes the expected (relative to p^i_{t+1}) payoff of player i; for an example, see Fudenberg and Levine (1998, Section 8.10).

23. Namely, such that the dependence of any one choice of $-i$ on any one past choice of i is small, relative to the number of periods; see Cahn (2000).

24. For a good test of the simplicity of a procedure, try to explain it verbally; in particular, consider the procedure of our Main Theorem vs. those requiring the computation of eigenvectors.

erty when discussing nonsophisticated behavior; this is the reason we have sought this result as our Main Theorem.

(g) *Inertia.* A specific and most distinctive feature by which the procedure of our Main Theorem differs from those of Theorem A and the other works mentioned above is that in the former the individual decisions privilege the most recent action taken: The probabilities used at period $t + 1$ are best thought of as propensities to depart from the play at t.

Viewed in this light, our procedure has significant inertial characteristics. In particular, there is a positive probability of moving from the strategy played at t only if there is another that appears better (in which case the probabilities of playing the better strategies are proportional to the regrets relative to the period t strategy).[25]

(h) *Friction.* The procedure (2.2) exhibits "friction": There is always a positive probability of continuing with the period t strategy.[26] To understand the role played by friction,[27] suppose that we were to modify the procedure (2.2) by requiring that the switching probabilities be rescaled in such a way that a switch occurs if and only if there is at least one better strategy (i.e., one with positive regret). Then the result of the Main Theorem may not hold. For example, in the familiar two-person 2×2 coordination game, if we start with an uncoordinated strategy pair, then the play alternates between the two uncoordinated pairs. However, no distribution concentrated on these two pairs is a correlated equilibrium.

It is worth emphasizing that in our result the breaking away from a bad cycle, like the one just described, is obtained not by ergodic arguments but by the probability of staying put (i.e., by friction). What matters is that the diagonal of the transition matrix be positive, rather than that all the entries be positive (which, indeed, will not hold in our case).

(i) *The set of correlated equilibria.* The set of correlated equilibria of a game is, in contrast to the set of Nash equilibria, geometrically simple: It is a convex set (actually, a convex polytope) of distributions. Since it includes

25. It is worth pointing out that if a player's last choice was j, then the relative probabilities of switching to k or to k' do not depend only on the average utilities that would have been obtained if j had been changed to k or to k' in the past, but also on the average utility that was obtained in those periods by playing j itself (it is the magnitude of the increases in moving from j to k or to k' that matters).

26. See Sanchirico (1996) and Section 4.6 in Fudenberg and Levine (1998) for a related point in a best-reply context.

27. See Step M7 in the Proof of the Main Theorem in the Appendix.

the Nash equilibria we know it is nonempty. Hart and Schmeidler (1989) (see also Nau and McCardle (1990)) provide an elementary (nonfixed point) proof of the nonemptiness of the set of correlated equilibria. This is done by using the Minimax Theorem. Specifically, Hart and Schmeidler proceed by associating to the given N-person game an auxiliary two-person zero-sum game. As it turns out, the correlated equilibria of the original game correspond to the maximin strategies of player I in the auxiliary game. More precisely, in the Hart-Schmeidler auxiliary game, player I chooses a distribution over N-tuples of actions, and player II chooses a pair of strategies for one of the N original players (interpreted as a play and a suggested deviation from it). The payoff to auxiliary player II is the expected gain of the designated original player if he were to follow the change suggested by auxiliary player II. In other words, it is the "regret" of that original player for not deviating. The starting point for our research was the observation that fictitious play applied to the Hart-Schmeidler auxiliary game must converge, by the result of Robinson (1951), and thus yield optimal strategies in the auxiliary game, in particular for player I—hence, correlated equilibria in the original game. A direct application of this idea does not, however, produce anything that is simple and separable across the N players (i.e., such that the choice of each player at time t is made independently of the other players' choices at t—an indispensable requirement).[28] Yet, our adaptive procedure is based on "no-regret" ideas motivated by this analysis and it is the direct descendant—several modifications later—of this line of research.[29]

(j) *The case of the unknown game.* The adaptive procedure of Section 2 can be modified[30] to yield convergence to correlated equilibria also in the case where players neither know the game, nor observe the choices of the other players.[31] Specifically, in choosing play probabilities at time $t + 1$, a

28. This needed "decoupling" across the N original players explains why applying linear programming-type methods to reach the convex polytope of correlated equilibria is not a fruitful approach. The resulting procedures operate in the space of N-tuples of strategies S (more precisely, in $\Delta(S)$), whereas adaptive procedures should be defined for each player i separately (i.e., on $\Delta(S^i)$).

29. For another interesting use of the auxiliary two-person zero-sum game, see Myerson (1997).

30. Following a suggestion of Dean Foster.

31. For similar constructions, see: Baños (1968), Megiddo (1980), Foster and Vohra (1993), Auer et al. (1995), Roth and Erev (1995), Erev and Roth (1998), Camerer and Ho (1998), Marimon (1996, Section 3.4), and Fudenberg and Levine (1998, Section 4.8). One may view this type of result in terms of "stimulus-response" decision behavior models.

player uses information *only* on his own actual past play and payoffs (and *not* on the payoffs that would have been obtained if his past play had been different). The construction is based on replacing $D_t^i(j, k)$ (see (2.1b)) by

$$C_t^i(j,k) := \frac{1}{t}\left[\sum_{\tau \le t : s_\tau^i = k} \frac{p_\tau^i(j)}{p_\tau^i(k)} u^i(s_\tau) - \sum_{\tau \le t : s_\tau^i = j} u^i(s_\tau) \right].$$

Thus, the payoff that player i would have received had he played k rather than j is estimated by the actual payoffs he obtained when he did play k in the past.

For precise formulations, results and proofs, as well as further discussions, the reader is referred to Hart and Mas-Colell (2000).

Center for Rationality and Interactive Decision Theory, Dept. of Economics, and Dept. of Mathematics, The Hebrew University of Jerusalem, Feldman Bldg., Givat-Ram, 91904 Jerusalem, Israel; hart@math.huji.ac.il; http://www.ma.huji.ac.il/~hart

and

Dept. d'Economia i Empresa, and CREI, Universitat Pompeu Fabra, Ramon Trias Fargas 25–27, 08005 Barcelona, Spain; mcolell@upf.es; http://www.econ.upf.es/crei/mcolell.htm

Manuscript received November, 1997; final revision received July, 1999.

Appendix: Proof of the Main Theorem

This appendix is devoted to the proof of the Main Theorem, stated in Section 2. The proof is inspired by the result of Section 3 (Theorem A). It is however more complex on account of our transition probabilities not being the invariant measures that, as we saw in Section 3, fitted so well with Blackwell's Approachability Theorem.

As in the standard proof of Blackwell's Approachability Theorem, the proof of our Main Theorem is based on a recursive formula for the distance of the vector of regrets to the negative orthant. However, our procedure (2.2) does not satisfy the Blackwell condition; it is rather a sort of iterative approximation to it. Thus, a simple one-period recursion (from t to $t + 1$) does not suffice, and we have to consider instead a multi-period recursion where a large "block" of periods, from t to $t + v$, is combined together. Both t and v are carefully chosen; in particular, t and v go to infinity, but v is relatively small compared to t.

We start by introducing some notation. Fix player i in N. For simplicity, we drop reference to the index i whenever this cannot cause confusion (thus we write D_t and R_t instead of D_t^i and R_t^i, and so on). Let $m := |S^i|$ be the number of strategies of player i, and let M be an upper bound on i's possible payoffs: $M \geq |u^i(s)|$ for all s in S. Denote $L := \{(j, k) \in S^i \times S^i : j \neq k\}$; then $\mathbb{R}^L$ is the $m(m-1)$-dimensional Euclidean space with coordinates indexed by L. For each $t = 1, 2, \ldots$ and each (j, k) in L, put[32]

$$A_t(j,k) = 1_{\{s_t^i = j\}}[u^i(k, s_t^{-i}) - u^i(s_t)],$$

$$D_t(j,k) = \frac{1}{t}\sum_{\tau=1}^{t} A_\tau(j,k),$$

$$R_t(j,k) = D_t^+(j,k) \equiv [D_t(j,k)]^+.$$

We shall write A_t for the vector $(A_t(j,k))_{j\neq k} \in \mathbb{R}^L$; the same goes for D_t, D_t^+, R_t, and so on. Let $\Pi_t(\cdot,\cdot)$ denote the transition probabilities from t to $t+1$ (these are computed after period t, based on h_t):

$$\Pi_t(j,k) := \begin{cases} \dfrac{1}{\mu} R_t(j,k), & \text{if } k \neq j, \\ 1 - \displaystyle\sum_{k' \neq j} \frac{1}{\mu} R_t(j,k'), & \text{if } k = j. \end{cases}$$

Thus, at time $t+1$ the strategy used by player i is to choose each $k \in S^i$ with probability $p_{t+1}^i(k) = \Pi_t(s_t^i, k)$. Note that the choice of μ guarantees that $\Pi_t(j,j) > 0$ for all $j \in S^i$ and all t. Finally, let

$$\rho_t := [\mathrm{dist}(D_t, \mathbb{R}_-^L)]^2$$

be the squared distance (in $\mathbb{R}^L$) of the vector D_t to the nonpositive orthant $\mathbb{R}_-^L$. Since the closest point to D_t in $\mathbb{R}_-^L$ is[33] D_t^-, we have $\rho_t = \|D_t - D_t^-\|^2 = \|D_t^+\|^2 = \Sigma_{j\neq k}[D_t^+(j,k)]^2$.

It will be convenient to use the standard "O" notation: For two real-valued functions $f(\cdot)$ and $g(\cdot)$ defined on a domain X, "$f(x) = O(g(x))$" means that there exists a constant $K < \infty$ such that $|f(x)| \leq Kg(x)$ for all x in[34] X. We write P for Probability, and E for Expectation. From now on, t, v,

32. We write 1_G for the indicator of the event G.

33. We write $[x]^-$ for $\min\{x, 0\}$, and D_t^- for the vector $([D_t(j,k)]^-)_{(j,k)\in L}$.

34. The domain X will usually be the set of positive integers, or the set of vectors whose coordinates are positive integers. Thus when we write, say, $f(t, v) = O(v)$, it means $|f(t, v)| \leq Kv$ for all v and t. The constants K will always depend only on the game (through N, m, M, and so on) and on the parameter μ.

and w will denote positive integers; $h_t = (s_\tau)_{\tau \le t}$ will be histories of length t; j, k, and s^i will be elements of S^i; s and s^{-i} will be elements of S and S^{-i}, respectively. Unless stated otherwise, all statements should be understood to hold "for all t, v, h_t, j, k, etc."; where histories h_t are concerned, only those that occur with positive probability are considered.

We divide the proof of the Main Theorem into 11 steps, M1–M11, which we now state formally; an intuitive guide follows.

- *Step M1:*

(i) $$E[(t+v)^2 \rho_{t+v} | h_t] \le t^2 \rho_t + 2t \sum_{w=1}^{v} R_t \cdot E[A_{t+w} | h_t] + O(v^2); \quad \text{and}$$

(ii) $$(t+v)^2 \rho_{t+v} - t^2 \rho_t = O(tv + v^2).$$

Define

$$\alpha_{t,w}(j, s^{-i}) := \sum_{k \in S^i} \Pi_t(k, j) P[s_{t+w} = (k, s^{-i}) | h_t] - P[s_{t+w} = (j, s^{-i}) | h_t].$$

- *Step M2:*

$$R_t \cdot E[A_{t+w} | h_t] = \mu \sum_{s^{-i} \in S^{-i}} \sum_{j \in S^i} \alpha_{t,w}(j, s^{-i}) u^i(j, s^{-i}).$$

- *Step M3:*

$$R_{t+v}(j,k) - R_t(j,k) = O\left(\frac{v}{t}\right).$$

For each $t > 0$ and each history h_t, define an auxiliary stochastic process $(\hat{s}_{t+w})_{w=0,1,2,\ldots}$ with values in S as follows: The initial value is $\hat{s}_t = s_t$, and the transition probabilities are[35]

$$P[\hat{s}_{t+w} = s | \hat{s}_t, \ldots, \hat{s}_{t+w-1}] := \prod_{i' \in N} \Pi_t^{i'}(\hat{s}_{t+w-1}^{i'}, s^{i'}).$$

(The $\hat{s}$-process is thus stationary: It uses the transition probabilities of period t at each period $t + w$, for all $w \ge 0$.)

- *Step M4:*

$$P[s_{t+w} = s | h_t] - P[\hat{s}_{t+w} = s | h_t] = O\left(\frac{w^2}{t}\right).$$

35. We write $\Pi_t^{i'}$ for the transition probability matrix of player i' (thus Π_t is Π_t^i).

Define

$$\hat{\alpha}_{t,w}(j, s^{-i}) := \sum_{k \in S^i} \Pi_t(k, j) P[\hat{s}_{t+w} = (k, s^{-i})|h_t] - P[\hat{s}_{t+w} = (j, s^{-i})|h_t].$$

- *Step M5:*

$$\alpha_{t,w}(j, s^{-i}) - \hat{\alpha}_{t,w}(j, s^{-i}) = O\left(\frac{w^2}{t}\right).$$

- *Step M6:*

$$\hat{\alpha}_{t,w}(j, s^{-i}) = P[\hat{s}^{-i}_{t+w} = s^{-i}|h_t][\Pi_t^{w+1} - \Pi_t^{w}](s_t^i, j),$$

where $\Pi_t^w \equiv (\Pi_t)^w$ is the wth power of the matrix Π_t, and $\left[\Pi_t^{w+1} - \Pi_t^w\right]$ (s_t^i, j) denotes the (s_t^i, j) element of the matrix $\Pi_t^{w+1} - \Pi_t^w$.

- *Step M7:*

$$\hat{\alpha}_{t,w}(j, s^{-1}) = O(w^{-1/2}).$$

- *Step M8:*

$$E[(t + v)^2 \rho_{t+v}|h_t] \leq t^2 \rho_t + O(v^3 + tv^{1/2}).$$

For each $n = 1, 2, \ldots$, let $t_n := [n^{5/3}]$ be the largest integer not exceeding $n^{5/3}$.

- *Step M9:*

$$E[t_{n+1}^2 \rho_{t_{n+1}}|h_{t_n}] \leq t_n^2 \rho_{t_n} + O(n^2).$$

- *Step M10:*

$$\lim_{n \to \infty} \rho_{t_n} = 0 \quad \text{a.s.}$$

- *Step M11:*

$$\lim_{t \to \infty} R_t(j, k) = 0 \quad \text{a.s.}$$

We now provide an intuitive guide to the proof. The first step (M1(i)) is our basic recursion equation. In Blackwell's Theorem, the middle term on the right-hand side vanishes (it is ≤ 0 by (3.2)). This is not so in our case; Steps M2–M8 are thus devoted to estimating this term. Step M2 yields an expression similar to (3.4), but here the coefficients α depend also on the moves of the other players. Indeed, given h_t, the choices s^i_{t+w} and s^{-i}_{t+w} are *not* independent when $w > 1$ (since the transition probabilities change with time). Therefore we replace the process $(s_{t+w})_{0 \leq w \leq v}$ by another process $\left(\hat{s}_{t+w}\right)_{0 \leq w \leq v}$, with a *stationary* transition matrix (that of period t). For w small relative to t, the change in probabilities is small (see Steps M3 and M4), and we estimate the total difference (Step M5). Next (Step M6), we factor out the moves of the other players (which, in the $\hat{s}$-process, are independent of the moves of player i) from the coefficients $\hat{\alpha}$. At this point we get the difference between the transition probabilities after w periods and after $w + 1$ periods (for comparison, in formula (3.4) we would replace both by the invariant distribution, so the difference vanishes). This difference is shown (Step M7) to be small, since w is large and the transition matrix has all its diagonal elements strictly positive.[36] Substituting in M1(i) yields the final recursive formula (Step M8). The proof is now completed (Steps M9–M11) by considering a carefully chosen subsequence of periods $(t_n)_{n=1,2,\ldots}$.

The rest of this Appendix contains the proofs of the Steps M1–M11.

- *Proof of Step M1:* Because $D_t^- \in \mathbb{R}^L_-$ we have

$$\rho_{t+v} \leq \left\|D_{t+v} - D_t^-\right\|^2 = \left\|\frac{t}{t+v} D_t + \frac{1}{t+v}\sum_{w=1}^{v} A_{t+w} - D_t^-\right\|^2$$

$$= \frac{t^2}{(t+v)^2}\left\|D_t - D_t^-\right\|^2 + \frac{2t}{(t+v)^2}\sum_{w=1}^{v}(A_{t+w} - D_t^-)\cdot(D_t - D_t^-)$$

$$+ \frac{v^2}{(t+v)^2}\left\|\frac{1}{v}\sum_{w=1}^{v} A_{t+w} - D_t^-\right\|^2$$

$$\leq \frac{t^2}{(t+v)^2}\rho_t + \frac{2t}{(t+v)^2}\sum_{w=1}^{v} A_{t+w}\cdot R_t + \frac{v^2}{(t+v)^2} m(m-1)16M^2.$$

Indeed: $|u^i(s)| \leq M$, so $|A_{t+w}(j,k)| \leq 2M$ and $|D_t(j,k)| \leq 2M$, yielding the upper bound on the third term. As for the second term, note that $R_t = D_t^+ = D_t - D_t^-$ and $D_t^- \cdot D_t^+ = 0$. This gives the bound of (ii). To get (i), take conditional expectation given the history h_t (so ρ_t and R_t are known). Q.E.D.

36. For further discussion on this point, see the Proof of Step M7.

• *Proof of Step M2:* We have

$$E[A_{t+w}(j,k)|h_t] = \sum_{s^{-i}} \phi(j,s^{-i})[u^i(k,s^{-i}) - u^i(j,s^{-i})],$$

where $\phi(j,s^{-i}) := P[s_{t+w} = (j,s^{-i})|h_t]$. So

$$R_t \cdot E[A_{t+w}|h_t] = \sum_j \sum_{k\neq j} R_t(j,k) \sum_{s^{-i}} \phi(j,s^{-i})[u^i(k,s^{-i}) - u^i(j,s^{-i})]$$

$$= \sum_{s^{-i}} \sum_j u^i(j,s^{-i}) \left[\sum_{k\neq j} R_t(k,j)\phi(k,s^{-i}) - \sum_{k\neq j} R_t(j,k)\phi(j,s^{-i}) \right]$$

(we have collected together all terms containing $u^i(j,s^{-i})$). Now, $R_t(k,j) = \mu\Pi_t(k,j)$ for $k \neq j$, and $\Sigma_{k\neq j} R_t(j,k) = \mu(1 - \Pi_t(j,j))$ by definition, so

$$R_t \cdot E[A_{t+w}|h_t] = \mu \sum_{s^{-i}} \sum_j u^i(j,s^{-i}) \left[\sum_k \Pi_t(k,j)\phi(k,s^{-i}) - \phi(j,s^{-i}) \right]$$

(note that the last sum is now over *all* k in S^i). *Q.E.D.*

• *Proof of Step M3:* This follows immediately from

$$(t+v)[D_{t+v}(j,k) - D_t(j,k)] = \sum_{w=1}^{v} A_{t+w}(j,k) - vD_t(j,k),$$

together with $|A_{t+w}(j,k)| \leq 2M$ and $|D_t(j,k)| \leq 2M$. *Q.E.D.*

• *Proof of Step M4:* We need the following Lemma, which gives bounds for the changes in the w-step transition probabilities as a function of changes in the 1-step transitions.

Lemma: Let $(X_n)_{n\geq 0}$ and $(Y_n)_{n\geq 0}$ be two stochastic processes with values in a finite set B. Assume $X_0 = Y_0$ and

$$|P[X_n = b_n|X_0 = b_0,\ldots,X_{n-1} = b_{n-1}]$$
$$- P[Y_n = b_n|Y_0 = b_0,\ldots,Y_{n-1} = b_{n-1}]| \leq \beta_n$$

for all $n \geq 1$ and all $b_0,\ldots,b_{n-1},b_n \in B$. Then

$$|P[X_{n+w} = b_{n+w}|X_0 = b_0,\ldots,X_{n-1} = b_{n-1}]$$
$$- P[Y_{n+w} = b_{n+w}|Y_0 = b_0,\ldots,Y_{n-1} = b_{n-1}]| \leq |B| \sum_{r=0}^{w} \beta_{n+r}$$

for all $n \geq 1$, $w \geq 0$, and all $b_0,\ldots,b_{n-1},b_{n+w} \in B$.

Proof: We write P_X and P_Y for the probabilities of the two processes $(X_n)_n$ and $(Y_n)_n$, respectively (thus $P_X[b_{n+w}|b_0,\ldots, b_{n-1}]$ stands for $P[X_{n+w} = b_{n+w}|X_0 = b_0,\ldots, X_{n-1} = b_{n-1}]$, and so on). The proof is by induction on w.

$$P_X[b_{n+w}|b_0,\ldots,b_{n-1}]$$
$$= \sum_{b_n} P_X[b_{n+w}|b_0,\ldots,b_n]P_X[b_n|b_0,\ldots,b_{n-1}]$$
$$\leq \sum_{b_n} P_Y[b_{n+w}|b_0,\ldots,b_n]P_X[b_n|b_0,\ldots,b_{n-1}] + |B|\sum_{r=1}^{w} \beta_{n+r}$$
$$\leq \sum_{b_n} P_Y[b_{n+w}|b_0,\ldots,b_n](P_Y[b_n|b_0,\ldots,b_{n-1}]+ \beta_n) + |B|\sum_{r=1}^{w} \beta_{n+r}$$
$$\leq P_Y[b_{n+w}|b_0,\ldots,b_{n-1}] + |B|\ \beta_n+ |B|\ \sum_{r=1}^{w} \beta_{n+r}$$

(the first inequality is by the induction hypothesis). Exchanging the roles of X and Y completes the proof. *Q.E.D.*

We proceed now with the proof of Step M4. From t to $t + w$ there are $|N|w$ transitions (at each period, think of the players moving one after the other, in some arbitrary order). Step M3 implies that each transition probability for the $\hat{s}$-process differs from the corresponding one for the s-process by at most $O(w/t)$, which yields, by the Lemma, a total difference of $|N|w|S|O(w/t) = O(w^2/t)$. *Q.E.D.*

- *Proof of Step M5:* Immediate by Step M4. *Q.E.D.*

- *Proof of Step M6:* Given h_t, the random variables $\left(\hat{s}^{i'}_{t+w}\right)_w$ are independent over the different players i' in N; indeed, the transition probabilities are all determined at time t, and the players randomize independently. Hence:

$$P[\hat{s}_{t+w} = (j, s^{-i})|h_t] = P[\hat{s}^{-i}_{t+w} = s^{-i}|h_t]P[\hat{s}^{i}_{t+w} = j|h_t],$$

implying that

$$\hat{\alpha}_{t,w}(j, s^{-i}) = P[\hat{s}^{-i}_{t+w} = s^{-i}|h_t]\left[\sum_{k\in S^i} \Pi_t(k, j)P[\hat{s}^{i}_{t+w} = k|h_t] - P[\hat{s}^{i}_{t+w} = j|h_t]\right].$$

Now $P[\hat{s}^i_{t+w} = j|h_t]$ is the probability of reaching j in w steps starting from s^i_t, using the transition probability matrix Π_t. Therefore $P[\hat{s}^i_{t+w} = j|h_t]$ is the (s^i_t, j)-element of the wth power $\Pi^w_t \equiv (\Pi_t)^w$ of Π_t, i.e., $[\Pi^w_t](s^i_t, j)$. Hence

$$\hat{\alpha}_{t,w}(j, s^{-i}) = P[\hat{s}^{-i}_{t+w} = s^{-i}|h_t]\left[\sum_{k\in S^i} \Pi_t(k, j)[\Pi^w_t](s^i_t, k) - [\Pi^w_t](s^i_t, j)\right]$$

$$= P[\hat{s}^{-i}_{t+w} = s^{-i}|h_t]\left[[\Pi^{w+1}_t](s^i_t, j) - [\Pi^w_t](s^i_t, j)\right],$$

completing the proof. *Q.E.D.*

• *Proof of Step M7:* It follows from M6 using the following Lemma (recall that $\Pi_t(j, j) > 0$ for all $j \in S^i$).

Lemma: Let Π be an $m \times m$ stochastic matrix with all of its diagonal entries positive. Then $[\Pi^{w+1} - \Pi^w](j, k) = O(w^{-1/2})$ for all $j, k = 1, \dots, m$.

Proof:[37] Let $\beta > 0$ be a lower bound on all the diagonal entries of Π, i.e., $\beta := \min_j \Pi(j, j)$. We can then write $\Pi = \beta I + (1 - \beta)\Lambda$, where Λ is also a stochastic matrix. Now

$$\Pi^w = \sum_{r=0}^{w} \binom{w}{r} \beta^{w-r}(1-\beta)^r \Lambda^r,$$

and similarly for Π^{w+1}. Subtracting yields

$$\Pi^{w+1} - \Pi^w = \sum_{r=0}^{w+1} \gamma_r \binom{w}{r} \beta^{w-r}(1-\beta)^r \Lambda^r,$$

where $\gamma_r := \beta(w + 1)/(w + 1 - r) - 1$. Now $\gamma_r > 0$ if $r > q := (w + 1)(1 - \beta)$, and $\gamma_r \le 0$ if $r \le q$; together with $0 \le \Lambda^r(j, k) \le 1$, we get

$$\sum_{r\le q} \gamma_r \binom{w}{r} \beta^{w-r}(1-\beta)^r \le [\Pi^{w+1} - \Pi^w](j,k) \le \sum_{r>q} \gamma_r \binom{w}{r} \beta^{w-r}(1-\beta)^r.$$

37. If Π were a strictly positive matrix, then $\Pi^{w+1} - \Pi^w \to 0$ would be a standard result, because then Π^w would converge to the invariant matrix. However, we know only that the diagonal elements are positive. This implies that, if w is large, then with high probability there will be a positive fraction of periods when the process does not move. But this number is random, so the probabilities of going from j to k in w steps or in $w + 1$ steps should be almost the same (since it is like having r "stay put" transitions versus $r + 1$).

Consider the left-most sum. It equals

$$\sum_{r \leq q} \binom{w+1}{r} \beta^{w+1-r}(1-\beta)^r - \sum_{r \leq q} \binom{w}{r} \beta^{w-r}(1-\beta)^r = G_{w+1}(q) - G_w(q),$$

where $G_n(\cdot)$ denotes the cumulative distribution function of a sum of n independent Bernoulli random variables, each one having the value 0 with probability β and the value 1 with probability $1-\beta$. Using the normal approximation yields (Φ denotes the standard normal cumulative distribution function):

$$G_{w+1}(q) - G_w(q) = \Phi(x) - \Phi(y) + O\left(\frac{1}{\sqrt{(w+1)}}\right) + O\left(\frac{1}{\sqrt{w}}\right),$$

where

$$x := \frac{q-(w+1)(1-\beta)}{\sqrt{(w+1)\beta(1-\beta)}} \quad \text{and} \quad y := \frac{q-w(1-\beta)}{\sqrt{w\beta(1-\beta)}};$$

the two error terms $O((w+1)^{-1/2})$ and $O(w^{-1/2})$ are given by the Berry-Esséen Theorem (see Feller (1965, Theorem XVI.5.1)). By definition of q we have $x = 0$ and $y = O(w^{-1/2})$. The derivative of Φ is bounded, so $\Phi(x) - \Phi(y) = O(x - y) = O(w^{-1/2})$. Altogether, the left-most sum is $O(w^{-1/2})$. A similar computation applies to the right-most sum. *Q.E.D.*

• *Proof of Step M8:* Steps M5 and M7 imply $\alpha_{t,w}(j, s^{-i}) = O(w^2/t + w^{-1/2})$. The formula of Step M2 then yields

$$R_t \cdot E[A_{t+w} \mid h_t] = O\left(\frac{w^2}{t} + w^{-1/2}\right).$$

Adding over $w = 1, 2, \ldots, v$ (note that $\sum_{w=1}^{v} w^\lambda = O(v^{\lambda+1})$ for $\lambda \neq -1$) and substituting into Step M1(i) gives the result. *Q.E.D.*

• *Proof of Step M9:* We use the inequality of Step M8 for $t = t_n$ and $v = t_{n+1} - t_n$. Because $v = [(n+1)^{5/3}] - [n^{5/3}] = O(n^{2/3})$, we have $v^3 = O(n^2)$ and $tv^{1/2} = O(n^{5/3+1/3}) = O(n^2)$, and the result follows. *Q.E.D.*

• *Proof of Step M10:* We use the following result (see Loève (1978, Theorem 32.1.E)):

Theorem (Strong Law of Large Numbers for Dependent Random Variables): *Let X_n be a sequence of random variables and b_n a sequence of real* *numbers increasing to* ∞, *such that the series* $\sum_{n=1}^{\infty} \operatorname{var}(X_n)/b_n^2$ *converges. Then*

$$\frac{1}{b_n}\sum_{\nu=1}^{n}\left[X_\nu - E[X_\nu|X_1,\ldots,X_{\nu-1}]\right] \xrightarrow[n\to\infty]{} 0 \quad a.s.$$

We take $b_n := t_n^2$, and $X_n := b_n\rho_{t_n} - b_{n-1}\,\rho_{t_{n-1}} = t_n^2\rho_{t_n} - t_{n-1}^2\,\rho_{t_{n-1}}$. By Step M1(ii) we have $|X_n| \leq O(t_n v_n + v_n^2) = O(n^{7/3})$, thus $\Sigma_n \operatorname{var}(x_n)/b_n^2 = \sum_n O(n^{14/3})/n^{20/3} = \sum_n O(1/n^2) < \infty$. Next, Step M9 implies

$$(1/b_n)\sum_{\nu\leq n} E[X_\nu|X_1,\ldots,X_{\nu-1}] \leq O\left(n^{-10/3}\sum_{\nu\leq n}\nu^2\right) = O(n^{-10/3}n^3) = O(n^{-1/3}) \to 0.$$

Applying the Theorem above thus yields that ρ_{t_n}, which is nonnegative and equals $(1/b_n)\Sigma_{\nu\leq n}X_\nu$, must converge to 0 a.s. *Q.E.D.*

• *Proof of Step M11:* Since $\rho_{t_n} = \sum_{j\neq k}[R_{t_n}(j,k)]^2$, the previous Step M10 implies that $R_{t_n}(j,k)\to 0$ a.s. $n\to\infty$, for all $j\neq k$. When $t_n \leq t \leq t_{n+1}$, we have $R_t(j,k) - R_{t_n}(j,k) = O(n^{-1})$ by the inequality of Step M3, so $R_t(j,k) \to 0$ a.s. $t\to\infty$. *Q.E.D.*

References

Auer, P., N. Cesa-Bianchi, Y. Freund, and R. E. Schapire (1995): "Gambling in a Rigged Casino: The Adversarial Multi-Armed Bandit Problem," in *Proceedings of the 36th Annual Symposium on Foundations of Computer Science,* 322–331.

Aumann, R. J. (1974): "Subjectivity and Correlation in Randomized Strategies," *Journal of Mathematical Economics,* 1, 67–96.

——— (1987): "Correlated Equilibrium as an Expression of Bayesian Rationality," *Econometrica,* 55, 1–18.

Baños, A. (1968): "On Pseudo-Games," *The Annals of Mathematical Statistics,* 39, 1932–1945.

Blackwell, D. (1956a): "An Analog of the Minmax Theorem for Vector Payoffs," *Pacific Journal of Mathematics,* 6, 1–8.

———(1956b): "Controlled Random Walks," in *Proceedings of the International Congress of Mathematicians 1954, Vol. III,* ed. by E. P. Noordhoff. Amsterdam: North-Holland, pp. 335–338.

Brown, G. W. (1951): "Iterative Solutions of Games by Fictitious Play," in *Activity Analysis of Production and Allocation,* Cowles Commission Monograph 13, ed. by T. C. Koopmans. New York: Wiley, pp. 374–376.

Cahn, A. (2000): "General Procedures Leading to Correlated Equilibria," The Hebrew University of Jerusalem, Center for Rationality DP-216.

Camerer, C., and T.-H. Ho (1998): "Experience-Weighted Attraction Learning in Coordination Games: Probability Rules, Heterogeneity, and Time-Variation," *Journal of Mathematical Psychology,* 42, 305–326.

Erev, I., and A. E. Roth (1998): "Predicting How People Play Games: Reinforcement Learning in Experimental Games with Unique, Mixed Strategy Equilibria," *American Economic Review,* 88, 848–881.

Feller, W. (1965): *An Introduction to Probability Theory and its Applications, Vol. II,* 2nd edition. New York: Wiley.

Foster, D., and R. V. Vohra (1993): "A Randomization Rule for Selecting Forecasts," *Operations Research,* 41, 704–709.

———(1997): "Calibrated Learning and Correlated Equilibrium," *Games and Economic Behavior,* 21, 40–55.

———(1998): "Asymptotic Calibration," *Biometrika,* 85, 379–390.

———(1999): "Regret in the On-line Decision Problem," *Games and Economic Behavior,* 29, 7–35.

Fudenberg, D., and D. K. Levine (1995): "Universal Consistency and Cautious Fictitious Play," *Journal of Economic Dynamics and Control,* 19, 1065–1089.

———(1998): *Theory of Learning in Games.* Cambridge, MA: The MIT Press.

———(1999): "Conditional Universal Consistency," *Games and Economic Behavior,* 29, 104–130.

Hannan, J. (1957): "Approximation to Bayes Risk in Repeated Play," in *Contributions to the Theory of Games, Vol. III,* Annals of Mathematics Studies 39, ed. by M. Dresher, A. W. Tucker, and P. Wolfe. Princeton: Princeton University Press, pp. 97–139.

Hart, S., and A. Mas-Colell (1999): "A General Class of Adaptive Strategies," The Hebrew University of Jerusalem, Center for Rationality DP-192, forthcoming in *Journal of Economic Theory.*

———(2000): "A Stimulus-Response Procedure Leading to Correlated Equilibrium," The Hebrew University of Jerusalem, Center for Rationality (mimeo).

Hart, S., and D. Schmeidler (1989): "Existence of Correlated Equilibria," *Mathematics of Operations Research,* 14, 18–25.

Loève, M. (1978): *Probability Theory, Vol. II,* 4th edition. Berlin: Springer-Verlag.

Luce, R. D., and H. Raiffa (1957): *Games and Decisions.* New York: Wiley.

Marimon, R. (1996): "Learning from Learning in Economics," in *Advances in Economic Theory,* ed. by D. Kreps. Cambridge: Cambridge University Press.

Megiddo, N. (1980): "On Repeated Games with Incomplete Information Played by Non-Bayesian Players," *International Journal of Game Theory,* 9, 157–167.

Mertens, J.-F., S. Sorin, and S. Zamir (1995): "Repeated Games, Part A," CORE DP-9420 (mimeo).

Myerson, R. B. (1997): "Dual Reduction and Elementary Games," *Games and Economic Behavior,* 21, 183–202.

Nau, R. F., and K. F. McCardle (1990): "Coherent Behavior in Noncooperative Games," *Journal of Economic Theory,* 50, 424–444.

Robinson, J. (1951): "An Iterative Method of Solving a Game," *Annals of Mathematics,* 54, 296–301.

Roth, A. E., and I. Erev (1995): "Learning in Extensive-Form Games: Experimental Data and Simple Dynamic Models in the Intermediate Term," *Games and Economic Behavior,* 8, 164–212.

Sanchirico, C. W. (1996): "A Probabilistic Model of Learning in Games," *Econometrica,* 64, 1375–1393.

— 10 —

Uncoupled Dynamics Do Not Lead to Nash Equilibrium

SERGIU HART
ANDREU MAS-COLELL*

It is notoriously difficult to formulate sensible adaptive dynamics that guarantee convergence to Nash equilibrium. In fact, short of variants of exhaustive search (deterministic or stochastic), there are no general results; of course, there are many important, interesting and well-studied particular cases. See the books of Jörgen W. Weibull (1995), Fernando Vega-Redondo (1996), Larry Samuelson (1997), Drew Fudenberg and David K. Levine (1998), Josef Hofbauer and Karl Sigmund (1998), H. Peyton Young (1998), and the discussion in Section IV below.

Here we provide a simple answer to the question: Why is that so? Our answer is that the lack of a general result is an intrinsic consequence of the natural requirement that dynamics of play be "uncoupled" among the players, that is, the adjustment of a player's strategy does not depend on the payoff functions (or utility functions) of the other players (it may depend on the other players' strategies, as well as on the payoff function of the player himself). This is a basic informational condition for dynamics of the "adaptive" or "behavioral" type.

American Economic Review 93 (2003), 1830–1836.

* Hart: Center for Rationality and Interactive Decision Theory, Department of Mathematics, and Department of Economics, The Hebrew University of Jerusalem, Feldman Building, Givat-Ram, 91904 Jerusalem, Israel (e-mail: hart@huji.ac.il; URL: <http://www.ma.huji.ac.il/~hart>); Mas-Colell: Department of Economics and Business, Universitat Pompeu Fabra, Ramon Trias Fargas 25–27, 08005 Barcelona, Spain (e-mail: mcolell@upf.es).
The research is partially supported by grants of the Israel Academy of Sciences and Humanities, the Spanish Ministry of Education, the Generalitat de Catalunya, and the EU-TMR Research Network. We thank Robert J. Aumann, Yaacov Bergman, Vincent Crawford, Josef Hofbauer, Piero La Mura, Eric Maskin, Motty Perry, Alexander Vasin, Bob Wilson, and the referees for their useful comments.

It is important to emphasize that, unlike the existing literature (see Section IV), we make no "rationality" assumptions: our dynamics are *not* best-reply dynamics, or better-reply, or payoff-improving, or monotonic, and so on. What we show is that the impossibility result is due only to an "informational" requirement—that the dynamics be uncoupled.

I. The Model

The setting is that of games in strategic (or normal) form. Such a game Γ is given by a finite set of players N, and, for each player $i \in N$, a strategy set S^i (not necessarily finite) and a payoff function[1] $u^i: \Pi_{j \in N} S^j \to \mathbb{R}$.

We examine differential dynamical systems defined on a convex domain X, which will be either $\Pi_{i \in N} S^i$ or[2] $\Pi_{i \in N} \Delta(S^i)$, and are of the form

$$\dot{x}(t) = F(x(t); \Gamma),$$

or $\dot{x} = F(x; \Gamma)$ for short. We also write this as $\dot{x}^i = F^i(x; \Gamma)$ for all i, where $x = (x^i)_{i \in N}$ and[3] $F = (F^i)_{i \in N}$.

From now on we keep N and $(S^i)_{i \in N}$ fixed, and identify a game Γ with its N-tuple of payoff functions $(u^i)_{i \in N}$, and a family of games with a set $\mathcal{U}$ of such N-tuples; the dynamics are thus

$$\dot{x}^i = F^i(x; (u^j)_{j \in N}) \text{ for all } i \in N. \tag{1}$$

We consider families of games $\mathcal{U}$ where every game $\Gamma \in \mathcal{U}$ has a single Nash equilibrium $\bar{x}(\Gamma)$. Such families are the most likely to allow for well-behaved dynamics. For example, the dynamic $\dot{x} = \bar{x}(\Gamma) - x$ will guarantee convergence to the Nash equilibrium starting from any initial condition.[4] Note, however, that in this dynamic $\dot{x}^i$ depends on $\bar{x}^i(\Gamma)$, which, in turn, depends on all the components of the game Γ, in particular on u^j for $j \neq i$. This motivates our next definition.

1. $\mathbb{R}$ denotes the real line.

2. We write $\Delta(A)$ for the set of probability measures over A.

3. For a well-studied example (see for instance Hofbauer and Sigmund, 1998), consider the class of "fictitious play"-like dynamics: the strategy $q^i(t)$ played by i at time t is some sort of "good reply" to the past play of the other players j, i.e., to the time average $x^j(t)$ of $q^j(\tau)$ for $\tau \leq t$; then (after rescaling the time axis) $\dot{x}^i = q^i - x^i \equiv G^i(x; \Gamma) - x^i \equiv F^i(x; \Gamma)$.

4. The same applies to various generalized Newton methods and fixed-point-convergent dynamics.

We call a dynamical system $F(x; \Gamma)$ (defined for Γ in a family of games $\mathcal{U}$) **uncoupled** if, for every player $i \in N$, the function F^i does *not* depend on u^j for $j \neq i$; i.e.,

$$\dot{x}^i = F^i(x; u^i) \text{ for all } i \in N \tag{2}$$

[compare with (1)]. Thus the change in player *i*'s strategy can be a function of the current N-tuple of strategies x and *i*'s payoff function u^i only.[5] In other words, if the payoff function of player i is identical in two games in the family, then at each x his strategy x^i will change in the same way.[6]

If, given a family $\mathcal{U}$ with the single-Nash-equilibrium property, the dynamical system always converges to the unique Nash equilibrium of the game for any game $\Gamma \in \mathcal{U}$—i.e., if $F(\bar{x}(\Gamma); \Gamma) = 0$ and $\lim_{t \to \infty} x(t) = \bar{x}(\Gamma)$ for any solution $x(t)$ (with any initial condition)—then we will call F a **Nash-convergent** dynamic for $\mathcal{U}$. To facilitate the analysis, we always restrict ourselves to C^1 functions F with the additional property that at the (unique) rest point $\bar{x}(\Gamma)$ the Jacobian matrix J of $F(\cdot\,; \Gamma)$ is hyperbolic and (asymptotically) stable—i.e., all eigenvalues of J have negative real parts.

We will show that:

There exist no uncoupled dynamics which guarantee Nash convergence.

Indeed, in the next two sections we present two simple families of games (each game having a single Nash equilibrium), for which uncoupledness and Nash convergence are mutually incompatible.

More precisely, in each of the two cases we exhibit a game Γ_0 and show that:[7]

Theorem 1: Let $\mathcal{U}$ be a family of games containing a neighborhood of the game Γ_0. Then every uncoupled dynamic for $\mathcal{U}$ is not Nash-convergent.

Thus an arbitrarily small neighborhood of Γ_0 is sufficient for the impossibility result (of course, nonexistence for a family $\mathcal{U}$ implies nonexistence for any larger family $\mathcal{U}' \supset \mathcal{U}$).

5. It may depend on the *function* $u^i(\cdot)$, not just on the current payoffs $u^i(x)$.

6. What the other players *do* (i.e., x^{-i}) is much easier to observe than *why* they do it (i.e., their utility functions u^{-i}).

7. An ε-*neighborhood* of a game $\Gamma_0 = (u_0^i)_{i \in N}$ consists of all games $\Gamma = (u^i)_{i \in N}$ satisfying $|u^i(s) - u_0^i(s)| < \varepsilon$ for all $s \in \Pi_{i \in N} S^i$ and all $i \in N$.

II. An Example with a Continuum of Strategies

Take $N = \{1, 2\}$ *and* $S^1 = S^2 = D$, where $D := \{z = (z_1, z_2) \in \mathbb{R}^2 : ||z|| \leq 1\}$ is the unit disk. Let $\phi : D \to D$ be a continuous function that satisfies:

- $\phi(z) = 2z$ for z in a neighborhood of 0; and
- $\phi(\phi(z)) \neq z$ for all $z \neq 0$.

Such a function clearly exists; for instance, let us put $\phi(z) = 2z$ for all $||z|| \leq 1/3$, define ϕ on the circle $||z|| = 1$ to be a rotation by, say, $\pi/4$, and interpolate linearly on rays between $||z|| = 1/3$ and $||z|| = 1$.

Define the game Γ_0 with payoff functions u_0^1 and u_0^2 given by[8]

$$u_0^i(x^i, x^j) := -\ ||x^i - \phi(x^j)||^2 \text{ for all } x^i, x^j \in D.$$

Γ_0 has a unique Nash equilibrium[9] $\bar{x} = (0, 0)$.

We embed Γ_0 in the family $\mathcal{U}_0$ consisting of all games $\Gamma \equiv (u^1, u^2)$ where, for each $i = 1, 2$, we have $u^i(x^i, x^j) = -||x^i - \xi^i(x^j)||^2$, with $\xi^i: D \to D$ a continuous function, such that the equation $\xi^i(\xi^j(x^i)) = x^i$ has a unique solution $\bar{x}^i$. Then $\bar{x} = (\bar{x}^1, \bar{x}^2)$ is the unique Nash equilibrium of the game[10] Γ.

We will now prove that every uncoupled dynamic for $\mathcal{U}_0$ is not Nash-convergent. This proof contains the essence of our argument, and the technical modifications needed for obtaining Theorem 1 are relegated to the Appendix. Let F thus be, by contradiction, a dynamic for the family $\mathcal{U}_0$ which is uncoupled and Nash-convergent. The dynamic can thus be written: $\dot{x}^i = F^i(x^i, x^j; u^i)$ for $i = 1, 2$.

The following key lemma uses uncoupledness repeatedly.

Lemma 2: Assume that y^i *is the unique* u^i*-best-reply of* i *to a given* y^j*, i.e.,* $u^i(y^i, y^j) > u^i(x^i, y^j)$ *for all* $x^i \neq y^i$*. Then* $F^i(y^i, y^j; u^i) = 0$*, and the eigenval-*

8. We use $j := 3 - i$ throughout this section. In the game Γ_0, each player i wants to choose x^i so as to match as closely as possible a function of the other player's choice, namely, $\phi(x^j)$.

9. $\bar{x}$ is a pure Nash equilibrium if and only if $\bar{x}^1 = \phi(\bar{x}^2)$ and $\bar{x}^2 = \phi(\bar{x}^1)$ or $\bar{x}^i = \phi(\phi(\bar{x}^i))$ for $i = 1, 2$. There are no mixed-strategy equilibria since the best reply of i to any mixed strategy of j is always unique and pure.

10. Moreover $\bar{x}$ is a strict equilibrium.

ues of the 2 × 2 *Jacobian matrix*[11] $J^i = \left(\partial F_k^i(y^i, y^j; u^i)/\partial x_l^i\right)_{k,l=1,2}$ *have negative real parts.*

Proof: Let Γ_1 be the game (u^i, u^j) with $u^j(x^i, x^j) := -\|x^j - y^j\|^2$ (i.e., ξ^j is the constant function $\xi^j(z) \equiv y^j$); then (y^i, y^j) is its unique Nash equilibrium, and thus $F^i(y^i, y^j; u^i) = 0$. Apply this to player j, to get $F^j(x^i, y^j; u^j) = 0$ for all x^i (since y^j is the unique u^j-best-reply to any x^i). Hence $\partial F_k^j(x^i, y^j; u^j)/\partial x_l^i = 0$ for $k, l = 1, 2$. The 4 × 4 Jacobian matrix J of $F(\cdot, \cdot; \Gamma_1)$ at (y^i, y^j) is therefore of the form

$$J = \begin{bmatrix} J^i & K \\ 0 & L \end{bmatrix}.$$

The eigenvalues of J—which all have negative real parts by assumption—consist of the eigenvalues of J^i together with the eigenvalues of L, and the result follows. ■

Put $f^i(x) := F^i(x; u_0^i)$; Lemma 2 implies that the eigenvalues of the 2 × 2 Jacobian matrix $J^i := \left(\partial f_k^i(0,0)/\partial x_l^i\right)_{k,l=1,2}$ have negative real parts. Again by Lemma 2, $f^i(\phi(x^j), x^j) = 0$ for all x^j, and therefore in particular $f^i(2x^j, x^j) = 0$ for all x^j in a neighborhood of 0. Differentiating and then evaluating at $\bar{x} = (0, 0)$ gives

$$2\partial f_k^i(0,0)/\partial x_l^i + \partial f_k^i(0,0)/\partial x_l^j = 0 \quad \text{for all } k, l = 1, 2.$$

Therefore the 4 × 4 Jacobian matrix J of the system (f^1, f^2) at $\bar{x} = (0, 0)$ is

$$J = \begin{bmatrix} J^1 & -2J^1 \\ -2J^2 & J^2 \end{bmatrix}.$$

Lemma 3: If the eigenvalues of J^1 and J^2 have negative real parts, then J has at least one eigenvalue with positive real part.

Proof: The coefficient a_3 of λ in the characteristic polynomial $\det(J - \lambda I)$ of J equals the negative of the sum of the four 3 × 3 principal minors; a straightforward computation shows that

$$a_3 = 3\det(J^1)\mathrm{trace}(J^2) + 3\det(J^2)\mathrm{trace}(J^1).$$

11. Subscripts denote coordinates: $x^i = (x_1^i, x_2^i)$ and $F^i = (F_1^i, F_2^i)$.

But $\det(J^i) > 0$ and $\text{trace}(J^i) < 0$ (since the eigenvalues of J^i have negative real parts), so that $a_3 < 0$.

Let $\lambda_1, \lambda_2, \lambda_3, \lambda_4$ be the eigenvalues of J. Then

$$\lambda_1\lambda_2\lambda_3 + \lambda_1\lambda_2\lambda_4 + \lambda_1\lambda_3\lambda_4 + \lambda_2\lambda_3\lambda_4$$
$$= -a_3 > 0$$

from which it follows that at least one λ_r must have positive real part. ■

This shows that the unique Nash equilibrium $\bar{x} = (0, 0)$ is unstable for $F(\,\cdot\,; \Gamma_0)$—a contradiction which establishes our claim.

For a suggestive illustration,[12] see Figure 1, which is drawn for x in a neighborhood of (0, 0) where $\phi(x^i) = 2x^i$. In the region $||x^2||/2 < ||x^1|| < 2||x^2||$ the dynamic leads "away" from (0, 0) (the arrows show that, for x^j fixed, the dynamic on x^i must converge to $x^i = 2x^j$—see Lemma 2).

III. An Example with Finitely Many Strategies

If the games have a finite number of strategies (i.e., if the S^i are finite), then the state space for the dynamics is the space of N-tuples of mixed strategies $\Pi_{i\in N}\, \Delta(S^i)$.

Consider a family $\mathcal{U}_0$ of three-player games where each player has two strategies, and the payoffs are:

0, 0, 0	a^1, 1, 0
1, 0, a^3	0, 1, a^3

0, a^2, 1	a^1, 0, 1
1, a^2, 0	0, 0, 0

where all the a^i are close to 1 (say, $1 - \varepsilon < a^i < 1 + \varepsilon$ for some small $\varepsilon > 0$), and, as usual, player 1 chooses the row, player 2 the column and player 3 the matrix.[13] Let Γ_0 be the game with $a^i = 1$ for all i; this game has been introduced by James Jordan (1993).

Denote by $x^1, x^2, x^3 \in [0, 1]$ the probability of the top row, left column, and left matrix, respectively. For every game in the family $\mathcal{U}_0$ there is a

12. The actual dynamic is 4-dimensional and may be quite complex.

13. Each player i wants to mismatch the next player $i + 1$, regardless of what player $i - 1$ does. (Of course, $i \pm 1$ is always taken modulo 3.)

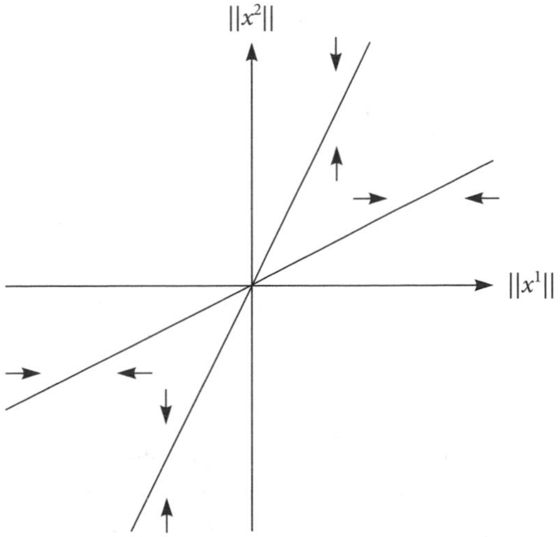

Figure 1. The Dynamic for the Game Γ_0 of Section II around (0, 0).

unique Nash equilibrium:[14] $\bar{x}^i(\Gamma) = a^{i-1}/(a^{i-1} + 1)$. In particular, for Jordan's game $\bar{x}_0 \equiv \bar{x}(\Gamma_0) = (1/2, 1/2, 1/2)$.

Let F be, by way of contradiction, an uncoupled Nash-convergent dynamic $\mathcal{U}_0$. For the game $\Gamma_0 \equiv \left(u_0^i\right)_{i=1,2,3}$ we denote $f^i(x^1, x^2, x^3) := F^i(x^1, x^2, x^3; u_0^i)$; let J be the 3×3 Jacobian matrix of f at $\bar{x}_0$.

For any y^1 (close to 1/2), the unique equilibrium of the game $\Gamma_1 = (u_0^1, u_0^2, u^3)$ in $\mathcal{U}_0$ with u^3 given by $a^3 = y^1/(1 - y^1)$ is $(y^1, 1/2, 1/2)$, and so $F^1(y^1, 1/2, 1/2; \Gamma_1) = 0$. This holds therefore also for Γ_0 since the dynamic is uncoupled: $f^1(y^1, 1/2, 1/2) = 0$ for all y^1 close to 1/2. Hence $\partial f^1(\bar{x}_0)/\partial x^1 = 0$. The same applies to the other two players, and we conclude that the diagonal—and thus the trace—of the Jacobian matrix J vanishes. Together with hyperbolicity [in fact, $\det(J) \neq 0$ suffices here], this implies the existence of an eigenvalue with positive real part,[15] thus establishing our contradiction—which proves Theorem 1 in this case.

14. In equilibrium: if i plays pure, then $i - 1$ plays pure, so all players play pure—but there is no pure equilibrium; if i plays completely mixed, then $a^i(1 - x^{i+1}) = x^{i+1}$.

15. Indeed: otherwise the real parts of all eigenvalues are 0. The dimension being odd implies that there must be a real eigenvalue. Therefore 0 is an eigenvalue—and the determinant vanishes.

We put on record that the uncoupledness of the dynamic implies additional structure on J. Indeed, we have $f^1(x^1, 1/2, x^3) = 0$ for all x^1 and x^3 close to 1/2 [since $(x^1, 1/2, x^3)$ is the unique Nash equilibrium when $a^1 = 1$—as in Γ_0—and $a^2 = x^3/(1 - x^3)$, $a^3 = x^1/(1 - x^1)$]. Therefore $\partial f^1(\bar{x}_0)/\partial x^3 = 0$ too, and so J is of the form

$$J = \begin{bmatrix} 0 & c & 0 \\ 0 & 0 & d \\ e & 0 & 0 \end{bmatrix}$$

for some real[16] c, d, e.

We conclude by observing that the specificities of the example have played very little role in the discussion. In particular, the property that the trace of the Jacobian matrix is null, or that f^i vanishes over a linear subspace of co-dimension 1, which is determined from the payoff function of player i only, will be true for any uncoupled dynamics at the equilibrium of a game with a completely mixed Nash equilibrium—provided, of course, that the game is embedded in an appropriate family of games.

IV. Discussion

(a) There exist uncoupled dynamics converging to *correlated equilibria*[17]—see Dean Foster and Rakesh V. Vohra (1997), Fudenberg and Levine (1999), Hart and Mas-Colell (2000),[18] and Hart and Mas-Colell (2003). It is thus interesting that Nash equilibrium, a notion that does not predicate coordinated behavior, cannot be guaranteed to be reached in an uncoupled way, while correlated equilibrium, a notion based on coordination, can.[19]

(b) In a general economic equilibrium framework, the parallel of Nash equilibrium is Walrasian (competitive) equilibrium. It is again well known that there are no dynamics that guarantee the general con-

16. If $cde \neq 0$ there is an eigenvalue with positive part, and if $cde = 0$ then 0 is the only eigenvalue.

17. Of course, these dynamics are defined on the appropriate state space of joint distributions $\Delta(\Pi_{i \in N} S^i)$, i.e., probability vectors on N-tuples of (pure) strategies.

18. In fact, the notion of "decoupling" appears in Section 4 (i) there.

19. *Cum grano salis* this may be called the "Coordination Conservation Law": there must be some coordination either in the equilibrium concept or in the dynamic.

vergence of prices to equilibrium prices if the dynamic has to satisfy natural uncoupledness-like conditions, for example, the nondependence of the adjustment of the price of one commodity on the conditions of the markets for other commodities (see Donald G. Saari and Carl P. Simon, 1978).

(c) In a mechanism-design framework, the counterpart of the uncoupledness condition is Leonid Hurwicz's "privacy-preserving" or "decentralized" condition—see Hurwicz (1986).

(d) There are various results in the literature, starting with Lloyd S. Shapley (1964, Sec. 5), showing that certain classes of dynamics cannot be Nash-convergent. These dynamics assume that the players adjust to the current state $x(t)$ in a way that is, roughly speaking, payoff-improving; this includes fictitious play, best-reply dynamics, better-reply dynamics, monotonic dynamics, adjustment dynamics, replicator dynamics, and so on; see Vincent P. Crawford (1985), Jordan (1993), Andrea Gaunesdorfer and Hofbauer (1995), Foster and Young (1998, 2001), and Hofbauer and Sigmund (1998, Theorem 8.6.1). All these dynamics are necessarily uncoupled (since a player's "good reply" to $x(t)$ depends only on his own payoff function). Our result shows that what underlies such impossibility results is not necessarily the rationality-type assumptions on the behavior of the players—but rather the informational requirement of uncoupledness.

(e) In a two-population evolutionary context, Alexander Vasin (1999) shows that dynamics that depend only on the vector of payoffs of each pure strategy against the current state—a special class of uncoupled dynamics—cannot be Nash-convergent.

(f) There exist uncoupled dynamics that are guaranteed to converge to (the set of) Nash equilibria for *special families* of games, like two-person zero-sum games, two-person potential games, dominance-solvable games, and others;[20] for some recent work see Hofbauer and William H. Sandholm (2002) and Hart and Mas-Colell (2003).

(g) There exist uncoupled dynamics that are most of the time close to

20. A special family of games may be thought of as giving information on the other players' payoff function (e.g., in two-person zero-sum games and potential games, u^i of one player determines u^j of the other player).

Nash equilibria, but are not Nash-convergent (they exit infinitely often any neighborhood of Nash equilibria); see Foster and Young (2002).

(h) Sufficient epistemic conditions for Nash equilibrium—see Robert J. Aumann and Adam Brandenburger (1995, Preliminary Observation, p. 1161)—are for each player i to know the N-tuple of strategies x and his own payoff function u^i. But that is precisely the information a player uses in an uncoupled dynamic—which we have shown *not* to yield Nash equilibrium. This points out the difference between the static and the dynamic frameworks: converging to equilibrium is a more stringent requirement than being in equilibrium.

(i) By their nature, differential equations allow strategies to be conditioned only on the limited information of the past captured by the state variable. It may thus be of interest to investigate the topic of this paper in more general settings.

Appendix

We show here how to modify the argument of Section II in order to prove Theorem 1. Consider a family of games $\mathcal{U}$ that is a neighborhood of Γ_0, and thus is certain to contain only those games in $\mathcal{U}_0$ that are close to Γ_0. The proof of Lemma 2 uses payoff functions of the form $u^j(x^i, x^j) = -||x^j - y^j||^2$ that do not depend on the other player's strategy x^i (i.e., ξ^j is the constant function $\xi^j(z) \equiv y^j$). Since the proof in Section II needs the result of Lemma 2 only for y^j in a neighborhood of 0, we will replace the above constant function ξ^j with a function that is constant in a neighborhood of the origin and is close to ϕ.

We will thus construct for each $a \in D$ with $||a|| < \varepsilon$ a function $\psi_a : D \to D$ such that: (1) $||\psi_a - \phi|| \leq C\varepsilon$ for some constant $C > 0$; (2) $\psi_a(z) = a$ for all $||z|| \leq 2\varepsilon$; (3) $\phi(\psi_a(z)) = z$ if and only if $z = \phi(a) = 2a$; and (4) $\psi_b(\psi_a(z)) = z$ if and only if $z = \psi_b(a) = b$. The games corresponding to (ϕ, ψ_{y^j}) and to (ψ_{x^i}, ψ_{y^j}), for $||y^j|| < \varepsilon/2$ and x^i close to $2y^j$, are therefore in $\mathcal{U}_0$ [by (3) and (4)], are close to Γ_0 [by (1)], and we can use them to obtain the result of Lemma 2 [by (2)].

The ψ functions may be constructed as follows: (i) $\psi_a(z) := a$ for $||z|| \leq 2\varepsilon$; (ii) $\psi_a(z) := 0$ for $||z|| = 3\varepsilon$; (iii) $\psi_a(z)$ is a rotation of $\phi(z)$ by the angle

ε for $||z|| \geq 4\varepsilon$; and (iv) interpolate linearly on rays in each one of the two regions $2\varepsilon < ||z|| < 3\varepsilon$ and $3\varepsilon < ||z|| < 4\varepsilon$.

It can be checked that conditions (1)–(4) above are indeed satisfied.[21]

References

Aumann, Robert J. and Brandenburger, Adam. "Epistemic Conditions for Nash Equilibrium." *Econometrica,* September 1995, *63*(5), pp. 1161–80.

Crawford, Vincent P. "Learning Behavior and Mixed-Strategy Nash Equilibrium." *Journal of Economic Behavior and Organization,* March 1985, *6*(1), pp. 69–78.

Foster, Dean and Vohra, Rakesh V. "Calibrated Learning and Correlated Equilibrium." *Games and Economic Behavior,* October 1997, *21*(1), pp. 40–55.

Foster, Dean and Young, H. Peyton. "On the Non-Convergence of Fictitious Play in Coordination Games." *Games and Economic Behavior,* October 1998, *25*(1), pp. 79–96.

———."On the Impossibility of Predicting the Behavior of Rational Agents." *Proceedings of the National Academy of Sciences,* October 2001, *98*(22), pp. 12848–53.

———. "Learning, Hypothesis Testing, and Nash Equilibrium." Mimeo, Johns Hopkins University, 2002.

Fudenberg, Drew and Levine, David K. *Theory of learning in games.* Cambridge, MA: MIT Press, 1998.

———. "Conditional Universal Consistency." *Games and Economic Behavior,* October 1999, *29*(1), pp. 104–30.

Gaunesdorfer, Andrea and Hofbauer, Josef. "Fictitious Play, Shapley Polygons, and the Replicator Equation." *Games and Economic Behavior,* November 1995, *11*(2), pp. 279–303.

Hart, Sergiu and Mas-Colell, Andreu. "A Simple Adaptive Procedure Leading to Correlated Equilibrium." *Econometrica,* September 2000, *68*(5), pp. 1127–50.

———. "Regret-Based Continuous-Time Dynamics." *Games and Economic Behavior,* January 2003, *45*(2), pp. 375–94.

Hofbauer, Josef and Sandholm, William H. "On the Global Convergence of Stochastic Fictitious Play." *Econometrica,* November 2002, *70*(6), pp. 2265–94.

21. (1) and (2) arc immediate. For (3), let $z = \phi(w)$ and $w = \psi_a(z)$. If $||z|| \leq 3\varepsilon$, then $w = \alpha a$ for some $\alpha \in [0, 1]$, therefore $||w|| < \varepsilon$ and $z = \phi(w) = 2w$, so in fact $||z|| < 2\varepsilon$ and $w = a$. If $||z|| > 3\varepsilon$, then, denoting by $\theta(x)$ the angle of x and recalling that ϕ rotates by an angle between 0 and $\pi/4$, we have $\theta(z) - \theta(w) = \theta(\phi(w)) - \theta(w) \in [0, \pi/4]$, whereas $\theta(w) - \theta(z) = \theta(\psi_a(z)) - \theta_z = \theta(\phi(z)) + \varepsilon - \theta(z) \in [\varepsilon, \pi/4 + \varepsilon]$, a contradiction. (4) is proven in a similar way.

Hofbauer, Josef and Sigmund, Karl. *Evolutionary games and population dynamics.* Cambridge: Cambridge University Press, 1998.

Hurwicz, Leonid. "Incentive Aspects of Decentralization," in K. J. Arrow and M. D. Intriligator, eds., *Handbook of mathematical economics,* Vol. III. Amsterdam: North-Holland, 1986, Ch. 28, pp. 1441–82.

Jordan, James. "Three Problems in Learning Mixed-Strategy Nash Equilibria." *Games and Economic Behavior,* July 1993, *5*(3), pp. 368–86.

Saari, Donald G. and Simon, Carl P. "Effective Price Mechanisms." *Econometrica,* September 1978, *46*(5), pp. 1097–125.

Samuelson, Larry. *Evolutionary games and equilibrium selection.* Cambridge, MA: MIT Press, 1997.

Shapley, Lloyd S. "Some Topics in Two-Person Games," in M. Dresher, L. S. Shapley, and A. W. Tucker, eds., *Advances in game theory* (*Annals of Mathematics Studies* 52). Princeton, NJ: Princeton University Press, 1964, pp. 1–28.

Vasin, Alexander. "On Stability of Mixed Equilibria." *Nonlinear Analysis,* December 1999, *38*(6), pp. 793–802.

Vega-Redondo, Fernando. *Evolution, games, and economic behavior.* New York: Oxford University Press, 1996.

Weibull, Jörgen W. *Evolutionary game theory.* Cambridge, MA: MIT Press, 1995.

Young, H. Peyton. *Individual strategy and social structure.* Princeton, NJ: Princeton University Press, 1998.

— APPENDIX A —

Rhyme and Reason

ANTONI BOSCH-DOMÈNECH
XAVIER CALSAMIGLIA
JOAQUIM SILVESTRE

Andreu Mas-Colell was born in Barcelona in 1944, the only child of Andreu Mas and Maria Colell. He began his undergraduate studies in economics at the *Universitat de Barcelona* in 1962, where he met Professor Manuel Sacristán, a philosopher and logician. In Andreu's own words, Sacristán was "one of the three or four strongest intellectual influences I have ever had . . . [From] him I inherited an unconditional faith on the power of rational analysis and an admiration for the scientific method."[1] Sacristán was a man of wide culture, deep convictions, and political passion. These traits also characterize Andreu's personality. During his college years Andreu was deeply committed to underground political activism against the Franco dictatorship. The regime caught up with him at some point and expelled him from the *Universitat de Barcelona.* Andreu and other Catalan students ended up completing their undergraduate degree in Bilbao, which was then formally a satellite campus of the Castilian *Universidad de Valladolid.*

The Spanish Navy drafted Andreu in 1966 and stationed him in its Madrid headquarters. This was a blessing in disguise, since it allowed him to participate in the graduate macroeconomics seminar of Professor Luis Ángel Rojo at the *Universidad Complutense de Madrid,* an oasis of liveliness in the bleak landscape of academic economics in Franco's Spain. Via Rojo's seminar, Andreu crossed the Atlantic and landed in Minnesota.

1. A. Mas-Colell, "Discurso pronunciado en el acto de recepción del Premio Rey Juan Carlos de Economía, instituido por la Fundación Celma Prieto," Bank of Spain (Madrid, 1989).

Minnesota

The Spanish connection to the Department of Economics at the University of Minnesota began with trips to Madrid by Minnesota businessman Dwayne Andreas and by Minnesota professor Walter W. Heller, former chair of the Council of Economic Advisers of John F. Kennedy. After meeting Professor Manuel Varela, a codirector of Rojo's seminar, Andreas decided to fund fellowships for Spanish graduate students at the Department of Economics in Minnesota. As Heller wrote in a department memo in 1967, "Andreas has a special interest in Spain, is trying to work with the liberalizing forces there on the commercial and financial sides, and if we were to decide that these Spanish students were really of good caliber, I'm quite sure he would accept our recommendation to provide financial support from his [Andreas] Foundation funds."[2]

Andreu and one of us (Antoni) were the first beneficiaries of the Andreas Fellowship. Andreu's political activism had interfered with his grades, and his undergraduate transcript did not impress the admission officials at the University of Minnesota, who asked Andreu's sponsors in Madrid (Rojo and Varela) whether Andreu's application was for real. They vouched for him: Minnesota took the risk and admitted Andreu. They never regretted it.

This was the start of a steady flow of students from both Barcelona and Madrid to the United States. So widespread was the influence of these Minnesota economists in Spain that the term "Minnesoto" came to mean any Spaniard who had completed a Ph.D. in economics at any United States university.

Once in Minneapolis, Andreu applied his capacities and energy to graduate work. The economics department was intellectually lively and a powerhouse for microeconomic theory and mathematical economics. This effort was led by Leo Hurwicz, John Chipman, and Ket Richter, with the younger additions of Hugo Sonnenschein and Thomas Muench, and attracted visiting professors Takashi Negishi and Hukukane Nikaido.

Andreu chose some classes in the mathematics department to complete his Ph.D. coursework requirements. In his second year, Andreu took the

2. Quoted by Tim Kehoe and Wendy Williamson, "The Spanish connection," Graduate Alumni Newsletter, Department of Economics, University of Minnesota (Fall 1997–Winter 1998).

Mathematical Economics course by Hugo Sonnenschein, whom Minnesota lost shortly afterward. Hugo's course was based on a list of ten open problems, and the students were expected to work in one of them. As mentioned by Hugo in his Introduction above, Andreu chose and solved one of the problems, resulting in a research paper that became the first of his publications, "General possibility theorems for group decisions," with Hugo (*Review of Economic Studies,* 1972).

Berkeley

The *Universitat Politècnica de Catalunya* hosted the European meetings of the Econometric Society, September 6–10, 1971. Andreu attended the presidential address by Gerard Debreu, where Debreu posed an open question on the approximation of arbitrary preferences by sequences of smooth ones, a question that could be addressed by differential topology methods. Andreu took the challenge and solved the problem in a few months, in what became his Ph.D. thesis.[3] Andreu's Ph.D. adviser, Ket Richter, apprised Debreu of Andreu's results, and Andreu was consequently offered a one-year appointment at Berkeley as an assistant research economist (financed by grants administered by Gerard Debreu and by Daniel MacFadden), later extended for two more years, followed by his appointment in 1975 to a joint tenure-track position in the departments of economics and mathematics. After only two years as an assistant professor and two more as an associate professor, he became professor in 1979.

During the 1970s Berkeley was a major center in mathematical economics: besides Debreu, we should mention Roy Radner in the economics department, David Gale in both the economics and the operations research departments, and John Harsanyi in the business school. Debreu had pioneered a new line of general equilibrium analysis grounded on differential topology, for which he enlisted the support of math professors Morris Hirsch and particularly Stephen Smale, plus the dedication of several younger researchers (Aloisio Araujo, Truman Bewley, Graciela Chichilnisky, Egbert and Hildegard Dierker, Birgit Grodal, Werner Hildenbrand, Wilhelm Neuefeind, Dieter Sondermann, and Karl Vind, among others). Andreu puts it thus: "As a young mathematical economist in Minnesota I saw

3. "Continuous and smooth consumers: Approximation theorems," *Journal of Economic Theory* 8(3) (1974): 305–336.

that gold had been discovered in California and I did not hesitate to join the rush."[4] Andreu quickly acquired a leading role in this collective effort.

Through the common friend Rose-Anne Dana (a student of Roy Radner), Andreu met Esther Silberstein, a Chilean graduate student in mathematics, working with Morris Hirsch, and an anti-Pinochet activist. They married in Oakland in 1976. Their first son, Alexandre (Alex), now an economics professor at Princeton, was born in Sacramento in 1977 and was followed by his siblings Gabriel (Boston, 1983), and Eva (Boston, 1985). Alex is the proud father of Anya and Clara. As we were checking the proofs of the present volume, Xavier, son of Gabriel and third grandchild of Andreu and Esther, was born in Manhattan.

Harvard

In 1981 Andreu was lured, with the enthusiastic support of Jerry Green, by Harvard University, where he spent fourteen years and became Louis Berkman Professor of Economics in 1988. There he took over the graduate microeconomics sequence, at first with Jerry Green and later, when Jerry became provost, with Michael Whinston. After several years of teaching the course, they had accumulated so much material that they decided to turn it into a textbook. Additional years of sweat and tears gave birth to the one and only graduate microeconomics textbook, so much so that in one social occasion in which one of us mentioned the name Mas-Colell, somebody whispered, Oh! A guy named after a book.

Andreu always kept close ties with Catalonia. He spent time in Barcelona every year and was a visiting professor at the *Universitat Autònoma de Barcelona* (UAB) in the academic year 1981–1982, where he taught graduate courses in game theory and welfare economics. Andreu has always been amazingly well informed about goings-on in his home country: he would subscribe to Spanish newspapers, keep his phone line constantly busy with transatlantic calls, and host visits by Spanish academicians, students, friends, friends of friends, and politicians. His appointment at Harvard afforded him a level of visibility and recognition that, as he soon discovered, could be marshaled to accomplish his long-held ambition to change the way things were done in Catalonia.

4. Quoted by A. Mas-Colell. "The determinacy of equilibria twenty-five years later," in *Economics in a Changing World: Proceedings of the Tenth World Congress of the International Economic Association, Moscow*, ed. B. Allen (Palgrave, 1996), 182–190.

These activities did not interfere with his intensely productive research, or with his editorial work. In addition, he served in 1989–1990 in the newly created position of assistant dean for affirmative action of the College of Arts and Sciences, in which capacity he started the personnel policy of stopping the tenure and promotion clocks for parenthood and for family caregiving.

His ability to focus in a theoretical challenge while ignoring his physical surroundings no doubt made all this possible. Andreu is an owl and likes to work at night. The American lifestyle forced him to have dinner no later than seven or eight in the evening, so he would start to feel hunger pangs in the wee hours of the night. Routinely, to assuage his cravings, he would place a steak on a hot grill and go back to his theorems while the meat was cooking, or rather, until the meat was burning and the shrill sound of the fire alarm reminded him, his family, and his guests that the meal was ready.

Berkeley was trying to lure Andreu back when he was there on sabbatical in 1985–1986, an option enhanced by Esther's contentment with returning to her teaching job at Mills College. Harvard counterattacked and adopted a devilish plan, attributed to Linda, the wife of Harvard econometrician Dale Jorgenson. One night Andreu and his family were at home in Berkeley, shortly after having finished dinner and getting the children ready for bed, when they heard a knock on their door. Unexpectedly, standing in the dark entrance, was Michael Spence. He was bringing a package-deal offer from Harvard for Andreu *and* Esther that they could not refuse.

It was at the Harvard Faculty Club that Andreu had dinner with Jordi Pujol, then president of the Catalan government, who was on an official visit to the East Coast. The chemistry between the two worked so well that Andreu was immediately called to supervise the creation of a new public university in Catalonia, the *Universitat Pompeu Fabra* (UPF) and to advise the Catalan government on several matters. This propitious meeting was the beginning of the end of Andreu's stay at Harvard. The call from his country and government turned out to be too strong to be resisted.

Barcelona

As mentioned, Andreu has kept a strong sense of responsibility for his country and a commitment to promote fundamental changes in Catalan academia and research. After the transition from the Franco dictatorship to a democratic regime, and especially after the first socialist electoral victory in 1982, Andreu's influence gathered momentum.

He was first involved in the creation of the research center *Institut d'Anàlisi Econòmica* (IAE), a part of the CSIC, a gigantic Spanish institution centrally administered.[5] Andreu learned from the CSIC experience that a top center must have a better control of hiring, promoting, and compensating personnel.

Andreu's second main project was the UPF. In 1990 he became a member of its first executive board and participated, from a distance but decisively, in its design. The UPF had to operate within the Spanish system but managed a better control of the selection and promotion processes, but not of the salary scale. This serious limitation induced Andreu to adopt a complementary approach, consisting in the creation of small, private-law institutions embedded in the public university system. These institutions have full control of the selection, compensation, and promotion of its personnel and yet operate in close cooperation with the universities. The CREI *(Centre de Recerca en Economia Internacional)* was the first step in that direction. It started in 1993 and has been working in close and fruitful collaboration with the UPF Department of Economics and Business since then.

Andreu left Harvard and moved to the UPF in 1995. When in 2000–2003 he became Minister of Universities, Research and the Information Society of the Catalan government, he had the opportunity to extend his influence and design a general institutional framework for science with three distinctive outcomes:

1. The creation of independent research centers in key areas of science, such as photonics, genomics, biomedicine, and chemistry, to name a few. Designed along the lines of CREI, these research centers are competing at the frontier of world research.
2. ICREA *(Institució Catalana de Recerca i Estudis Avançats),* a flexible and independent foundation, which recruits top scientists and makes them available to interested research institutions: this makes ICREA a science hotbed "without walls." Today, more than 230 top world researchers fill ICREA's permanent positions in Catalan universities and research centers. To give an idea of its impact, ICREA

5. The acronym CSIC stands for *Consejo Superior de Investigaciones Científicas.* The IAE is located in the campus of the *Universitat Autònoma de Barcelona.*

has obtained more European Research Grants than has the whole Spanish university system.

3. The Serra Húnter Program, which recruits faculty members for Catalan public universities following standards of international excellence, with a flexible and internationalized contractual system that avoids the rigidity of the civil servant model prevalent in Spain.

During his first tenure as minister, Andreu's abilities at persuasion provided a big impulse toward the establishment in Catalonia of large research infrastructures, such as the Barcelona Supercomputing Center and the Synchrotron Radiation Facility.

Andreu was also the creator of the Barcelona Graduate School of Economics (GSE) and became, in 2006, the first chairman of its board. The GSE is a joint undertaking of the UPF Department of Economics and Business, the UAB Unit of Economic Analysis, the IAE, and the CREI, offering world-class graduate education. It has become one of the leading clusters of economics research in Europe.

In 2010, Andreu was again appointed minister of the Catalan government, this time of economy and knowledge, in the middle of a serious economic and budgetary crisis.

All this activity has not deterred Andreu from pushing his research at full throttle, in no small measure due to his relation with Sergiu Hart (Hebrew University Rationality Center, Jerusalem), a relation that started with Sergiu's visit to Harvard in 1984 and evolved into a deep and enduring friendship.[6] Sergiu and Andreu have collaborated for about twenty years, with Andreu visiting Jerusalem and Sergiu visiting Harvard or Barcelona several times a year. Their tight cooperation has delivered extremely elegant papers, the Catalan government unremitting budgetary crisis notwithstanding.

Building upon the shoulders of the mathematical economics pioneers, Andreu Mas-Colell has moved the frontiers of knowledge and changed the way economics is taught. At the same time, a continuous and smooth thread, a political passion, links his early student activism with his current responsibilities in the Catalan government. Knowing, feeling, and doing in harmony: this is Andreu, rhyme and reason.

6. See S. Hart and A. Mas-Colell, *Simple Adaptive Strategies: From Regret-Matching to Uncoupled Dynamics*, World Scientific Series in Economic Theory, vol. 4 (Hackensack, New Jersey: World Scientific, 2013).

—— APPENDIX B ——

List of Scientific Publications

Books and Pamphlets

Noncooperative Approaches to the Theory of Perfect Competition (editor), Academic Press, 1982. It includes a Presentation chapter.

The Theory of General Economic Equilibrium: A Differentiable Approach, Cambridge University Press, 1985. Paperback edition, 1990. Spanish translation by A. Manresa published by Fundación Argentaria, 1992.

Contributions to Mathematical Economics, in Honor of Gérard Debreu, (joint editor with W. Hildenbrand), North-Holland, 1986.

Equilibrium Theory and Applications (joint editor with W. A. Barnett, B. Cornet, C. d'Aspremont and J. J. Gabszewicz), Cambridge University Press, 1991.

Microeconomic Theory (with M. Whinston and J. Green), Oxford University Press, 1995. Chinese translation: 2001.

Cooperation: Game Theoretic Approaches (joint editor with Sergiu Hart) NATO ASI Series. Series F, Vol. 155, Springer Verlag, 1997.

Nuevas Fronteras de la Política Económica, 1998, (joint editor with Massimo Motta, published by CREI, 1999).

Higher Aspirations: an Agenda for Reforming European Universities (with Philippe Aghion, Mathias Dewatripont, Caroline Hoxby, and André Sapir). Bruegel Blueprint Series No. 5, Bruegel, 2008.

Simple Adaptive Strategies, From Regret-Matching to Uncoupled Dynamics (with S. Hart). In *World Scientific Series in Economic Theory,* Vol. 4, 2013.

Translation into Spanish of G. Debreu's *Theory of Value* (with J. Oliu). Editorial Bosch, Barcelona, 1974.

Articles

1. "General possibility theorems for group decisions" (with H. Sonnenschein). *Review of Economic Studies* 39(2): 205–212, 1972.
2. "Racionalidad y decisividad en la teoría de la elección social." *Moneda y Crédito* 122: 3–12, 1972.
3. "A Model of intersectoral migration and growth" (with A. Razin). *Oxford Economic Papers* 25(1): 72–79, 1973.
4. "Instantaneous and noninstantaneous adjustment in two-sector growth models" (with A. Bosch and A. Razin). *Metroeconomica* 25: 1–14, 1973.
5. "La regla de oro de la acumulación y la elasticidad de sustitución unitaria en modelos bisectoriales." *Revista Española de Economía* 3(2): 75–95, 1973.
6. "Continuous and smooth consumers: Approximation theorems." *Journal of Economic Theory* 8(3): 305–336, 1974.
7. "A Note on a theorem of F. Browder." *Mathematical Programming* 6(1): 229–233, 1974.
8. "A Characterization of community excess demand functions" (with D. McFadden, R. Mantel, and M. Richter). *Journal of Economic Theory* 9(4): 361–374, 1974.
9. "An equilibrium existence theorem without complete or transitive preferences." *Journal of Mathematical Economics* 1(3): 237–246, 1974. (Chapter 1 in this volume.)
10. "Algunas observaciones sobre la teoría del tatonnement de Walras en economías productivas." *Anales de Economía* 21, 1974.
11. Introduction to *Elección Social y Valores Individuales* (Spanish Translation of K. Arrow, *Social Choice and Individual Values*). Madrid: Instituto de Estudios Fiscales, 1974.
12. "A further result on the representation of games by markets." *Journal of Economic Theory* 10: 117–122, 1975.
13. "On the continuity of equilibrium prices in constant returns production economies." *Journal of Mathematical Economics* 2(1): 21–33, 1975.
14. "An equilibrium existence theorem for a general model without ordered preferences" (with D. Gale). *Journal of Mathematical Economics* 2(1): 9–15, 1975. Also, "Corrections to 'An equilibrium existence

theorem for a general model without ordered preferences.'" *Journal of Mathematical Economics* 6(3): 297–298, 1979.

15. "A model of equilibrium with differentiated commodities." *Journal of Mathematical Economics* 2(2): 263–295, 1975. (Chapter 2 in this volume.)
16. "En torno a una propiedad poco atractiva del equilibrio competitivo." *Moneda y Crédito* 136: 11–27, 1976.
17. "A remark on a smoothness property of convex preorders." *Journal of Mathematical Economics* 3(1): 103–105, 1976.
18. "The demand theory of the weak axiom of revealed preference" (with R. Kihlstrom and H. Sonnenschein). *Econometrica* 44(5): 971–978, 1976.
19. "The recoverability of consumer's preferences from market demand behavior." *Econometrica* 45(6): 1409–1430, 1977.
20. "Sobre la computatión de los equilibrios en economías lineales." *Cuadernos de Economía* 5(14): 443–461, 1977.
21. "Some generic properties of excess demand and an application" (with W. Neufeind). *Econometrica* 45(3): 591–600, 1977.
22. "Regular non-convex economies." *Econometrica* 45(6): 1387–1407, 1977.
23. "Indivisible commodities and general equilibrium theory." *Journal of Economic Theory* 12(2): 443–456, 1977.
24. "On the continuous representation of preorders." *International Economic Review* 18(2): 509–513, 1977.
25. "Competitive and value allocations of large exchange economies." *Journal of Economic Theory* 14(2): 419–438, 1977.
26. "On the equilibrium price set of a pure exchange economy." *Journal of Mathematical Economics* 4(2): 117–126, 1977. (Chapter 3 in this volume.)
27. "On revealed preference analysis." *Review of Economic Studies* 45(1): 121–131, 1978.
28. "On the role of complete transitive preferences in equilibrium theory" (with D. Gale). In G. Schwodiauer, ed., *Equilibrium and Disequilibrium in Economic Theory,* Reidel, 1978.
29. "Notes on the smoothing of aggregate demand" (with A. Araujo). *Journal of Mathematical Economics* 5(2): 113–127, 1978.
30. "A note on the Core Equivalence Theorem: How many blocking co-

alitions are there?" *Journal of Mathematical Economics* 5(3): 207–215, 1978.

31. "A selection theorem for open graph correspondences with star-shaped values." *Journal of Mathematical Analysis and Applications* 68(1): 273–275, 1979.
32. "Homeomorphisms of compact, convex sets and the Jacobian matrix." *SIAM Journal of Mathematical Analysis* 10(6): 1105–1109, 1979.
33. "Two propositions on the global univalence of systems of cost functions." In J. Green and J. Scheinkman, eds., *General Equilibrium Growth and Trade*, Academic Press, 1979.
34. "A refinement of the Core Equivalence Theorem." *Economic Letters* 3(4): 307–310, 1979.
35. "Remarks on the game-theoretic analysis of a simple distribution of surplus problems." *International Journal of Game Theory* 9(3): 125–140, 1980.
36. "Efficiency properties of strategic market games: An axiomatic approach" (with P. Dubey and M. Shubik). *Journal of Economic Theory* 22(2): 339–362, 1980.
37. "Efficiency and decentralization in the pure theory of public goods." *Quarterly Journal of Economics* 94(4): 625–641, 1980. (Chapter 4 in this volume.)
38. "Revealed preference after Samuelson." In G. Feiwel, ed., *Samuelson and Neo-Classical Economics*, Martin Neijoff, 1981.
39. "Perfect competition and the Core." *Review of Economic Studies* 49(1): 15–30, 1982.
40. "The Cournotian foundations of Walrasian equilibrium theory: An Exposition of Recent Theory." In W. Hildenbrand, ed., *Advances in Economic Theory,* Econometric Society Publication No. 1, Cambridge University Press, 1981.
41. "Walrasian equilibria as limits of non-cooperative equilibria: Mixed strategies." *Journal of Economic Theory* 30(1): 153–170, 1983.
42. "Teoría del desempleo en Keynes y en la actualidad." *Información Comercial Española* 593: 68, 1983.
43. "Gestion au coût marginal et efficacité de la production agrégée: un exemple" (with P. Beato). *Annales de l'INSEE* 51: 39–46, 1983.
44. "An observation on gross substitutability and the Weak Axiom of Revealed Preference" (with T. Kehoe). *Economic Letters* 15(3–4): 241–243, 1984.

45. "The marginal cost pricing rule as a regulation mechanism in mixed markets" (with P. Beato). In M. Marchand, P. Pestieau, and H. Tulkens, eds., *The Performance of Public Enterprises,* North-Holland, 1984.
46. "The profit motive in the theory of monopolistic competition." *Zeitschrift für Nationalokonomie/Journal of Economics,* Supplementum 4, 1984.
47. "On a theorem of Schmeidler." *Journal of Mathematical Economics* 13(3): 201–206, 1984.
48. "On marginal cost pricing with given tax-subsidy rules" (with P. Beato). *Journal of Economic Theory* 37(2): 356–365, 1985.
49. "On the theory of perfect competition." *1984 Nancy L. Schwarz Memorial Lecture delivered at the Kellogg School of Management, Northwestern University.* The Kellog School, 1985.
50. "Pareto optima and equilibria: The finite dimensional case." In C. Aliprantis, O. Burkinshaw, and N. Rothman, eds., *Advances in Equilibrium Theory,* Lecture Notes in Economics, No. 244, Springer-Verlag, 1985.
51. "La libre entrada y la eficiencia económica: Un análisis de equilibrio parcial." *Revista de Economía Española* 2(1): 135–152, 1985.
52. "An introduction to the differentiable approach in the theory of economic equilibrium." In S. Reiter, ed., *Studies in Mathematical Economics,* Vol. 25 of the AMS Studies in Mathematics, American Mathematical Society, 1986.
53. "Notes on price-quantity tatonnement dynamics." In H. Sonnenschein, ed., *Models of Economics Dynamics,* Lecture Notes in Economics and Mathematical Systems 264: 49–68, Springer-Verlag, 1986.
54. "Valuation equilibrium and Pareto optimum revisited." Chapter 17 in A. Mas-Colell and W. Hildenbrand, eds., *Advances in Mathematical Economics,* North-Holland, 1986.
55. "Four lectures on the differentiable approach to general equilibrium theory." In A. Ambrosetti, F. Gori, and R. Lucchetti, eds., *Mathematical Economics (Montecatini Terme),* Lecture Notes in Mathematics, No. 1330, Springer-Verlag, 1986.
56. "The price equilibrium existence problem in topological vector lattices." *Econometrica* 54(5): 1039–1054, 1986. (Chapter 5 in this volume.)

57. "Engel curves leading to the weak axiom in the aggregate" (with X. Freixas). *Econometrica* 55(3): 515–531, 1987.
58. "On the second welfare theorem for anonymous net trades in exchange economies with many agents." In T. Groves, R. Radner, and S. Reiter, eds., *Information, Incentives and Economic Mechanisms: Essays in Honor of L. Hurwicz,* University of Minnesota Press, p. 267–292, 1987.
59. "Economías con rendimientos crecientes a escala." German Bernácer Lectures 1987, Universidad de Alicante, published by The Generalitat Valenciana, 1987.
60. "Algunos comentarios sobre la teoría cooperativa de los juegos." *Cuadernos Económicos* 40(3): 143–162, 1988.
61. "An example of core allocations and Pareto optima far from agents' demand correspondences" (with R. Anderson). Appendix to R. Anderson: "The second welfare theorem with nonconvex preferences," *Econometrica* 56(2): 361–382, 1988.
62. "The potential of the Shapley Value" (with S. Hart). In A. E. Roth, ed., *The Shapley Value*, Cambridge University Press, p. 127, 1988.
63. "Real indeterminacy with financial assets" (with J. Geanakoplos). *Journal of Economic Theory* 47(1): 22–38, 1989. (Chapter 6 in this volume.)
64. "Potential, value and consistency" (with S. Hart). *Econometrica* 57(3): 589–614, 1989. (Chapter 7 in this volume.)
65. "Cost share equilibria: A Lindahlian approach" (with J. Silvestre). *Journal of Economic Theory* 47(2): 239–256, 1989.
66. "Recent developments in the theory of incomplete markets." *Economic Record* 65, Issue 188: 82–84, 1989.
67. "Capital theory paradoxes: Anything goes." Chapter 17 in G. Feiwel, ed., *Joan Robinson and Modern Economic Theory*, New York: New York University Press; distributed by Columbia University Press, p. 505–520, 1989.
68. "An equivalence theorem for a bargaining set." *Journal of Mathematical Economics* 18(2): 129–139, 1989. (Chapter 8 in this volume.)
69. "Determinacy of equilibrium in large-square economies" (with T. Kehoe, D. Levine, and W. Zame). *Journal of Mathematical Economics* 18(3): 231–262, 1989.
70. "A geometric approach to a class of fixed point theorems" (with

M. C. Hirsch and A. Magill). *Journal of Mathematical Economics* 19(1–2): 95–106, 1990.

71. "A new approach to the existence of equilibria in vector lattices" (with S. Richard). *Journal of Economic Theory* 53(1): 1–11, 1991.
72. "Sobre el problema del determinisme en teoria econòmica." In X. Calsamiglia and R. Marimon, eds., *Invitació a la Teoria Econòmica,* Barcelona: Ediciones Ariel, 1991.
73. "Gross substitutability in large square economies" (with D. Levine, T. Kehoe and M. Woodford). *Journal of Economic Theory* 54(1): 1–25, 1991.
74. "On the uniqueness of equilibrium once again." In W. A. Barnett, B. Cornet, C. d'Aspremont, J. Gabszewicz, and A. Mas-Colell, eds., *Equilibrium Theory and Applications,* Cambridge University Press, 1991.
75. "On the finiteness of the number of critical equilibria, with an application to random selections" (with J. Nachbar). *Journal of Mathematical Economics* 20(4): 397–419, 1991.
76. "Equilibrium theory in infinite dimensional spaces (with W. Zame). In W. Hildenbrand and Sonnenschein, eds., *Handbook of Mathematical Economics,* Vol. 4, North Holland, p. 1835–1898, 1991.
77. "Indeterminacy in incomplete market economies." *Economic Theory* 1(1): 45–61, 1991.
78. "An Axiomatic approach to the efficiency of non-cooperative equilibrium in economies with a continuum of traders." In M. A. Khan and N. C. Yannelis, eds., *Equilibrium Theory in Infinite Dimensional Spaces,* Springer-Verlag, 1991.
79. "A Note on cost-share equilibrium and consumer-owners" (with J. Silvestre). *Journal of Economic Theory* 54(1): 204–214, 1991.
80. "Infinite-dimensional equilibrium theory: Discussion of Jones." In J. J. Laffont, ed., *Advances in Economic Theory: Sixth World Congress,* Vol. 2, Cambridge University Press, 1992.
81. "Three observations on sunspots and asset redundancy." In P. Dasgupta, D. Gale, O. Hart, and E. Maskin, eds., *Economic Analysis of Markets and Games,* MIT Press, 1992.
82. "Equilibrium theory with possibly satiated preferences." In Mukul Majumdar, ed., *Equilibrium and Dynamics: Essays in Honor of David Gale,"* New York: St. Martin's Press, p. 201–213, 1992.

83. "On aggregate demand in a measure space of agents," CMRS Working Paper IP-183, University of California, Berkeley, 1973. Published in W. Neuefeind and R. G. Riezman, eds., *Economic Theory and International Trade: Essays in Memoriam, J. Trout Rader*, Springer-Verlag, 1992.
84. "A noncooperative interpretation of Value and Potential" (with S. Hart). In R. Selten, ed., *Rational Interaction: Essays in Honor of John C. Harsanyi*, Springer-Verlag, 1992.
85. "Implementation in economies with a continuum of agents" (with X. Vives). *Review of Economic Studies* 60(203): 613–629, 1993.
86. "Elogi del creixement econòmic." In Jordi Nadal, ed., *El món cap a on anem*, Eumo Editorial, p. 183–214, 1994.
87. "Sobre el caràcter obligatori i universal de l'assegurança de salut." In G. López Casasnovas, ed., *Anàlisi Econòmica de la Sanitat*, Departament de Sanitat, Generalitat de Catalunya, p. 63–75, 1994.
88. "L'anàlisi econòmica i el dret." *Iuris: Quaderns de Política Jurídica* 3: 73–85, 1994.
89. Two entries for the New Palgrave Dictionary of Economics: "Cooperative Equilibria" and "Non-Convexities," 1994.
90. "Stationary Markov equilibria" (with D. Duffie, J. Geanakoplos, and A. MacLennan). *Econometrica* 62(4): 745–782, 1994.
91. "Egalitarian solutions of large games: I. A continuum of players" (with S. Hart). *Mathematics of Operations Research* 20(4): 959–1002, 1995.
92. "Egalitarian solutions of large games: II. The asymptotic approach" (with S. Hart). *Mathematics of Operations Research* 20(4): 1003–1022, 1995.
93. "The determinacy of equilibria twenty-five years later." *Economics in a Changing World*, Proceedings of the tenth World Congress of the *International Economic Association*, p. 182–190, Moscow, 1992 (B. Allen, ed.) Palgrave 1996.
94. "Self-fulfilling equilibria: An existence theorem for a general state space" (with P. K. Monteiro). *Journal of Mathematical Economics* 26(1): 51–62, 1996.
95. "The existence of security market equilibrium with a non-atomic state space" (with W. Zame). *Journal of Mathematical Economics* 26(1): 63–84, 1996.
96. "Harsanyi values of large economies: Nonequivalence to competitive

equilibria" (with S. Hart). *Games and Economic Behavior* 13(1): 74–99, 1996.

97. "Bargaining and Value" (with S. Hart). *Econometrica* 64(2): 357–380, 1996.
98. "Bargaining games." In S. Hart and A. Mas-Colell, eds., *Cooperation: Game-Theoretic Approaches,* Springer, p. 69–91, 1996.
99. "On the stability of best reply and gradient systems with applications to imperfectly competitive models" (with L. Corchon). *Economic Letters* 51(1): 59–65, 1996.
100. "Finite horizon bargaining and the consistent field" (with A. Gomes and S. Hart). *Games and Economic Behavior* 27(2): 204–228, 1999.
101. "The future of general equilibrium." *Spanish Economic Review* 1(3): 207–214, 1999.
102. "Should cultural goods be treated differently?" *Journal of Cultural Economics,* Special Issue, Barcelona Conference Plenary Papers 23(1–2): 87–93, 1999.
103. "A simple adaptive procedure leading to Correlated Equilibrium" (with S. Hart). *Econometrica* 68(5): 1127–1250, 2000. (Chapter 9 in this volume.)
104. "A note on the decomposition (at a point) of aggregate excess demand on the Grassmanian" (with P. Gottardi). *Journal of Mathematical Economics* 33(4): 463–473, 2000.
105. "A general class of adaptive strategies" (with S. Hart). *Journal of Economic Theory* 98(1): 26–54, 2001.
106. "A reinforcement procedure leading to Correlated Equilibrium" (with S. Hart). In G. Debreu, W. Neuefeind, and W. Trockel, eds., *Economic Essays: A Festschrift for Werner Hildenbrand,* Springer, p. 181–200, 2001.
107. "Uncoupled dynamics do not lead to Nash Equilibrium" (with S. Hart). *American Economic Review* 93(5): 1830–1836, 2003. (Chapter 10 in this volume.)
108. "Regret-based continuous-time dynamics" (with S. Hart). *Games and Economic Behavior* 45(2): 375–394, 2003.
109. "The European space of higher education: Incentive and governance issues. *Rivista di Politica Economica* 93(11–12): 11–30, 2003.
110. "La estructura institucional y la financiación de la educación superior: Una aproximación basada en la teoría del capital humano." *Políticas, Mercados e Instituciones Económicas,* Vol. I (José Pérez,

Carlos Sebastián y Pedro Tedde, eds.) *Editorial Complutense*, p. 525–543, 2004.

111. "Birgit Grodal: A friend to her friends." *Journal of the European Economic Association* 2(5): 906–912, 2004. Reprinted in C. Schultz and K. Vind, eds., *Institutions, Equilibria and Efficiency, Essays in Honor of Birgit Grodal*, Springer 2006.
112. "Stochastic uncoupled dynamics and Nash Equilibrium" (with S. Hart). *Games and Economic Behavior* 57(2): 286–303, 2006.
113. "Notes on stochastic choice." In C. Aliprantis, R. Matzkin, D. McFadden, J. Moore, and N. Yannelis, eds., *Rationality and Equilibrium. A Symposium in Honor of Marcel K. Richter*, Springer, p. 243–252, 2006.
114. "Nash Equilibrium and economics: Remarks." *Computer Science Review* 1(2): 100–102, 2007.
115. "Why reform Europe's universities?" (with P. Aghion, M. Dewatripont, C. Hoxby, and A. Sapir). *Bruegel Policy Brief*, Issue 2007/04, 2007.
116. "The governance and performance of universities: Evidence from Europe and the US" (with P. Aghion, M. Dewatripont, C. Hoxby, and A. Sapir). *Economic Policy* 25, Issue 61, 2009.
117. "Bargaining and cooperation in strategic form games" (with S. Hart). *Journal of the European Economic Association* 8(1): 7–33, 2010.
118. "Generic finiteness of equilibrium in bimatrix games." *Journal of Mathematical Economics* 46: 382–383, 2010.
119. "Markets, correlation and regret matching" (with S. Hart). *Games and Economic Behavior*, forthcoming 2016.